Introduction to Graphic Communication

by

Harvey Robert Levenson, Ph. D.
Department Head
Graphic Communication Department
California Polytechnic State University
San Luis Obispo, California

The Good Neighbor Press & Services
P. O. Box 323
Atascadero, California 93423
Phone: 805-466-3745
Fax: 805-466-8770
E-mail: gnps@thegrid.net

Dedication

This book is dedicated to all of those who promote graphic communication as the most meaningful, detailed, comprehensive, and informative form of communication that has ever existed and that may ever exist.

Preface

Introduction to Graphic Communication was written to address a need for a broad-based publication dealing with graphic communication background, concepts, technologies, processes, segments, and products. This is a survey book covering the graphic communication industry including traditional and digital printing, and non-print imaging.

The book is a "snapshot" of an industry's past, present, and prospects for the future. It is meant to broadly acquaint the reader with the issues and processes of graphic communication. It is not an all-encompassing treatise, but it will allow the reader to intelligently understand the industry.

The book covers new printing and imaging technology issues impacting day-to-day communication of people in all fields of study and occupations—graphic communication has become nearly everyone's business. Regardless of a person's field or discipline, graphic communication is practiced everywhere.

Introduction to Graphic Communication has a section on the historical growth and role of graphic communication in society, and even one on the history of the Internet and the World Wide Web. The book goes on to focus on providing a practical knowledge of the industry and its processes—traditional and electronic, print and non-print digital imaging.

This version of *Introduction to Graphic Communication* is a prelude to a more comprehensive one soon to be published. It will include an extensive glossary of approximately 8,000 terms bridging traditional and digital graphic communication, and computer terms. It will be the first book that recognizes that printing and computers have merged to redefine the graphic arts industry, and to really understand this industry one must know the terms common to both fields.

Introduction to Graphic Communication was written for people new to the field. It is for students studying the field at a two-year college, four-year college, and at other schools having graphic communication as a program of study. On the professional level, it is for those entering the field in the areas of sales, marketing, production, technology, graphic design, advertising, and print buying. It is also for those new to the equipment manufacturing and vendor segments of the industry. The book is concise and to the point.

The book covers graphic communication background, industry structure, processes, consumable supplies, technologies, markets, segments, and products. The focus is on traditional and digital reproduction technology. The relationships between prepress imaging and press, and post press operations are explored. A survey of prepress operations include art and copy preparation, desktop publishing, photomechanical and electrophotographic techniques, imagesetting, image manipulation, and image carriers. A survey of imaging and press operations include lithography, gravure, flexography, screen printing, letterpress, and non-impact printing such as electrostatic and inkjet printing, and on-demand digital imaging and printing. A survey of post press operations include finishing and related processes for the delivery and distribution of print and related digital media.

Consumable supplies including substrates, ink, and toners are also covered. There are special chapters on DRUPA, the world's largest graphic communication industry exposition, and on sustainability by contributing author, Don Carli.

I thank all of the vendors that have contributed images and information to this book. All are credited. I express particular thanks to Océ Printing Systems for access to images from the company's book, *Digital Printing*, and for printing promotional copies of *Introduction to Graphic Communication* at Print 05 in Chicago. A particular thank you is extended to Duncan Newton and Denis Williams of Océ for their efforts and support in producing this book.

Finally, I thank Tamara Moore, professional editor, who edited *Introduction to Graphic Communication.* Tamara is also Managing Editor, Scientific Publications Carnegie Museum of Natural History.

I hope that this book is useful in a practical sense and insightful to students and industry professionals alike.

Harvey Robert Levenson, Ph. D.

Table of Contents

1

What Is Graphic Communication?

Graphic communication is not an art or a science—It is a combination of both. It involves creative skills as well as technical standards. While it was once considered a trade or a craft focused solely on printing, graphic communication today is an industry. A trade that required redundant tasks to create products has given way to a number of diverse services requiring high-level management and professional preparation.

Traditionally, graphic communication has been divided into three categories: prepress, press, and post-press. However, with the broadening of the definition to mean more than printing alone, there's a need to redefine the categories that encompass the broad field of graphic communication. Considering the introduction of non-print media under the umbrella of graphic communication—such as Web authoring, Internet publishing, and related imaging areas—more appropriate categories to define the field are design reproduction technology, electronic publishing and imaging, and printing and imaging management, and even packaging graphics has benefited from advances in imaging.

Design Reproduction Technology

Design reproduction technology links design and technology, but it is not the study or practice of graphic design. It involves ensuring that a design is produced and then manipulated in ways that are optimal for its application. Repurposing is an important part of this process. Repurposing means taking a particular piece of artwork and converting it to suit the medium for which it is going to be used. For example, the various technical characteristics of a piece of artwork will be different if the artwork is going to be used for a print brochure than if it will be displayed on a Web site.

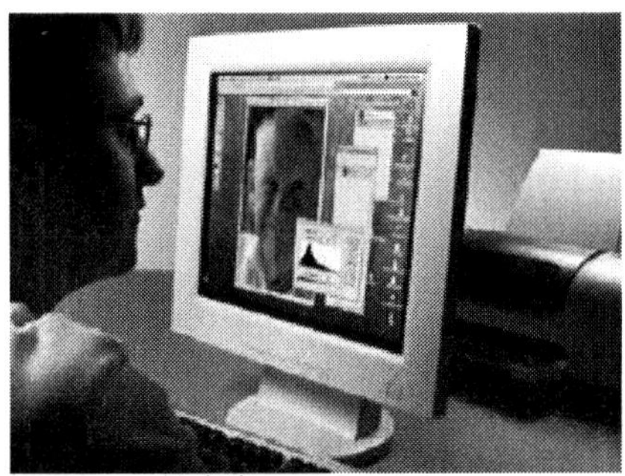

Image manipulation on a screen. (Kodak Polychrome Graphics)

The technical characteristics for print and computer screens or television monitors are very different. They also differ for printing on standard paper as opposed to inkjet photo paper, or burning to a CD versus printing out a photograph to hang on a wall. The difference related to things such as resolution, screen rulings, color gamut, and more. Design reproduction technology takes into account such considerations before an image is moved to the medium on which it will ultimately appear.

Design reproduction technology links art and technology. It frees the graphic designer to focus on the aesthetic and creative facets of design rather than on the technology, which has become the job of the design reproduction technologist. This design technologist understands the language and role of the graphic designer as well as the roles of those involved in reproducing design in print and distributing the designs electronically. The design reproduction technologist understands how to take art and copy and transform them into a form that will look good on a monitor or as a printed piece.

Design reproduction technologists require an understanding of design principles as well as a high level of computer skills. They either develop a design for the client, or the client gives them the design and they make it production-ready for print or electronic media. Hence, it is vital that design reproduction technologists be familiar with the different types of printing processes as well as the processes involved in creating Web sites, DVDs, CD-ROMS, and related media.

Today design and artwork are produced on computers using software. Software packages are available for all major operating platforms such as Windows and Macintosh. Design reproduction technologists must be versed in each major platform, though Macintosh presently dominates the graphic arts industry. These platforms and related software are used for color and type selection, sizing, electronic retouching, pagination, and more. The work of the design reproduction technologist in making images ready for production and reproduction can lead to high-resolution hard proofs produced on color printers or to images on a computer monitor for soft proofing or final imaging. In some cases the finished image can go directly to a printing plate or directly to a printing press cylinder.

In any event, the final work of the design reproduction technologist must precisely meet the expectations of the client. This is achieved by working closely with the creative people involved in the design phase of the project and with the technical staff involved in its final reproduction or distribution.

Employment prospects for qualified design reproduction technologists are very strong. The most dramatic growth in employment opportunities over the next decade will be in the areas of digital printing and in non-print electronic imaging.

In sum, the design reproduction technologist focuses on "graphic thinking" and understanding the principles of design and technology as they relate to advertising and publication production, packaging, commercial printing, and other media. Additionally, the focus is on manipulating designs through the use of digital cameras and scanners and preparing designs for line and halftone media using copy manipulation techniques and computer applications. The design technologist also understands the technical limitations and production requirements for print and non-print production and distribution.

Advertising agencies, publishing companies, and any other industry segments that combine art, design, and technology in producing a final product engage the services of a design reproduction technologist

Summary of Design Reproduction Technology

• Links art with technology.
• Understands the language and job functions of the graphic designer and the imaging technologist.
• Knows how to transform art and copy into a form that will look good on a monitor or as a printed piece.

It focuses on:

- Graphic "thinking"
- Design technology and principles
- Advertising and publication production
- Pre-separated art for cameras and scanners
- Line and halftone media
- Modern copy technology
- Technical limitations and production requirements
- Computerized design and copy preparation

Electronic Publishing and Imaging

Electronic publishing and imaging addresses issues related to hardware and software development and application. The electronic publishing and imaging specialist is versed in the growing array of equipment and peripherals used in creating print and non-print media. This specialist is also aware of graphic communication software applications and how to optimize the use of such software. Some may even be involved in software development or hold positions with software and hardware corporations in product development and testing.

Electronic publishing and imaging specialists understand digital imaging preparation and manipulation. They know the technical procedures for coordinating the red, green, and blue of color monitors with the yellow, cyan, and magenta of pigments of full-color printing on paper, among other things that help coordinate the color space between monitor images and printed sheets.

The breadth of electronic publishing and imaging areas include electronic controls, computer hardware and software, integrated systems, computer-aided typesetting and imagesetting, new technologies, digital imaging and media,

Internet applications, Web authoring, interactive documents, multimedia, imaging systems management, file management, and research and development.

Electronic publishing and imaging specialists typically have a degree in graphic communication, graphic arts, or a related computer science field. Graphic communication software and hardware undergo major upgrades every two to three years and these specialists must keep up with trends and developments.

Press sheet imposition on a computer. (Kodak Polychrome Graphics)

Today nearly every company associated with the graphic communication industry can benefit from having an electronic publishing and imaging specialist on its staff. This includes printers, publishers, packaging companies, advertising agencies, as well as hardware and software development companies. Such specialists have become particularly valuable for companies that have entered the area of digital printing and variable data printing where the development and maintenance of data files are particularly important.

Summary of Electronic Publishing and Imaging

- Understands digital image preparation and manipulation.
- Prepares for positions in software and hardware corporations in product development and testing.
- Knows the technical procedures for coordinating the red, green, and blue of color monitors with yellow, cyan, and magenta of pigments printed on paper.

It focuses on:

- Electronic controls
- Computer hardware and software
- Integrated systems
- Computer-aided typesetting and imagesetting
- New technologies
- Digital imaging and media
- Internet applications
- Interactive documents
- Multimedia
- Imaging systems management
- File management
- Research and development

Printing and Imaging Management

Printing and imaging management includes all facets of running companies involved in new and traditional media. This includes all segments of the printing industry such as prepress, press, and post press for commercial printing, publication printing, packaging, and related areas. It also includes management issues faced by equipment and supply vendors as well software and hardware developers for the graphic arts. The printing and imaging management specialist knows how to relate budget limitations to achieving desired results, as well as how to manage and train personnel to produce high-quality work efficiently. These specialists understand planning, management and systems analysis, quality control, and production control. They are versed in price estimating and financial controls, marketing and sales strategies, personnel relations, statistical process control (SPC), and total quality management (TQM).

Printing and imaging management professionals enjoy working with people and financial figures, and have a firm grasp of the various business aspects of running and growing a company. Through such management involvement, they provide products and services needed by customers of printing and imaging companies.

Some of the most important aspects of printing and imaging businesses are the products and services that flow to the customer. They must be produced at the lowest possible price, at the highest possible quality, as quickly as possible. Effective management helps achieve this.

Another important aspect of management in the graphic arts is cash flow analysis and ensuring that there is sufficient cash coming into the company to cover all expenses, overhead, salaries, equipment, and supplies, plus a reasonable profit. Marketing and sales are crucial to this and to the overall success of printing and imaging companies. Knowing how to market and sell in the graphic arts is particularly important in light of new digital technologies that provide printing on-demand and variable data products. Selling the new technology of printing and imaging takes special skills.

Effective printing and imaging management involves setting and carrying out goals; understanding appropriate workflows and digital file management; and being able to make the right decisions quickly. Perhaps most important, it involves motivating people and communicating effectively and thoughtfully.

Summary of Printing and Imaging Management

• Covers all facets of running companies involved in traditional and new media.
• Involves understanding how to relate budget limitations to achieving desired outcomes.

• Emphasizes the importance of managing and training personnel to produce high-quality work efficiently.

It focuses on:
• Plant management
• Planning
• Manufacturing and systems analysis
• Quality control
• Production control
• Estimating sand financial controls
• Marketing and sales
• Personnel relations
• Statistical process control (SPC)
• Total quality management (TQM)

Packaging Graphics
Packaging has been an enormous part of the graphic arts for decades and continues to be one of the fastest growing segments of the graphic communication industry. It is the only segment that has not been negatively impacted by the Internet and the World Wide Web, and in fact has grown.

Packaging graphics is a diverse field emphasizing digital file creation, technology, and printing. The field also involves an understanding of structural design and food packaging as they relate to consumer and industrial print packaging. Other important facets of packaging graphics are marketing, digital graphic creation, and print reproduction processes combined with inline and offline converting processes.

Evaluation of wine label color. (Kodak Polychrome Graphics)

The packaging field enjoys huge annual sales through its five sub-segments: folding carton packaging; flexible packaging; metal decorating; label printing; and corrugated board printing. Most facets of packaging graphics require some of the highest quality control standards in all of the graphic arts because of the need to reproduce colors that look natural—such as food, flesh tones, wood grains, and more.

This segment of the graphic arts industry faces environmental concerns related to non-biodegradable packages and issues of recycling packaging materials. Government and concerned citizens routinely exert pressure to counteract the proliferation of packages that are not recyclable and, hence, add to landfill disposal sites.

An interesting trend in the field has to do with the growth of short-run package printing with the use of digital printing presses. Nearly every major producer of digital presses is developing technology designed specifically for package printing.

The packaging graphics field is highly creative and requires people who can help product manufacturers stay competitive through unique graphics that portray the products contained within. There is stringent competition among such companies. It is often package design and structure that will make or break a company.

Summary of Packaging Graphics

• Expects continued growth.

• Requires an understanding of creativity and technology that serve consumer interests.

• Deals with issues of environmental protection and recyclables.

It focuses on:

• Package design and creativity
• Package substrates
• Technology
• Environment
• Package engineering
• Consumer tastes
• Competition

2

History of Printing

Overview

Printing as an industry dates back to 1456 with Johann Gutenberg's invention of the process of duplicating movable type. Gutenberg made possible the mass production of printing for the first time. Before Gutenberg, books and other documents were laboriously produced by hand. Using a modified wine press, Gutenberg automated the process and tremendously spread the process of producing and disseminating printing. Today, the printing industry still has the goal of mass-producing and distributing documents as quickly as possible.

Prior to 1970 change was the exception, not the rule in the printing industry. Little changed in process or technology over long periods of time. For example, Gutenberg's invention lived on as the main method for printing until 1886, when Ottmar Mergenthaler invented the Linotype machine. That was a 430-year span with little change in the main printing processes. During this time span, three-color printing was invented in 1710; the halftone process was invented in 1871; and the first four-color rotary printing press was developed in 1892. The application of halftones today is the same as the application of halftones then, and the application of four-color printing today is the same as it was then.

The offset lithographic press came about in the early 1900s and introduced the concept of using rubber blankets as a mechanism to transfer an image from a printing plate to paper. The concept of using rubber blankets for transferring images continues to be central to the offset lithographic process today.

The linotype machine lived on as the main technology for producing type for printing until 1954 when the Harris Corporation invented the phototypesetting machine. This allowed setting type on photographic film or paper as opposed to slugs of metal type produced by the linotype machine. Phototypesetting them became the main method of producing text and numbers for printing until the late 1980s when desktop publishing became a practical reality for producing first text and then graphics for printing. Even then, this represented a spread of about 35 years during which a basic technology was used for the purposes of producing printing. Desktop publishing has grown in use and sophistication and is rapidly replacing traditional prepress in the printing industry.

Other technologies that have impacted how printing is produced include computer graphics, integrated systems for color separations and color control, on-demand variable data printing, and Internet publishing. However, the very fundamental premise of what printing is and the concepts of producing images using line art, halftones, screens, and full-color process printing has remained the same in some cases for decades and in some cases for centuries.

Detailed

The story of civilization is the story of communications that leads to the story of printing. No institutions such as education, law, religion, or medicine would exist in their present state had it not been for printing.

All culture in western societies is a culture of the printed word. Printing has had an astonishing influence on human history. The printing press is one of the three inventions that have had the greatest influence on the world, the other two being gunpowder and the compass. Two individuals who influenced the world more than any others are Ts'ai Lun, who invented the process for making paper in China in 105 A.D., and Johann Gutenberg. In many studies conducted at the turn of the latest millennium regarding the most important invention of the previous 1000 years, printing and movable type were ranked number one more than any other invention.

Johann Gutenberg
(Biografias y Vidas)

Ts'ai Lun's and Gutenberg's inventions altered the structure of institutions and influenced how learning took place. They defined printing as the most meaningful, detailed, pervasive, and informative form of mass communication. Additionally, their inventions lived on for centuries. Ts'ai Lun's "pulping" process for making paper is still used today on highly automated machines costing tens of millions of dollars. Gutenberg's invention survived for nearly 500 years before it was replaced by the Linotype machine and then by photographic and then computerized typesetting technology. The Chinese are credited with inventing movable type two thousand years before Gutenberg; Gutenberg showed how to duplicate movable type.

Printing developed around the world because of technology transfer resulting from transportation, commerce, and wars. One speculation is that printing is an extension of human evolution and the seemingly innate need for humans to recall, keep records, and communicate.

The invention of printing and its subsequent development marks a turning point in the history of civilization. It has changed views about literary art and style and modified the psychological processes by which words are used for communication. The invention of printing broadcast the printed language and gave to print a degree of authority that it has never lost. Printing was the first public address system and soon became a new form of expression. This accomplishment, for printing, involved the invention of paper as well as inks made with an oil base, the development of engraving on woodblocks, and the development of the printing press and the special techniques of presswork involved in printing.

An oil base for printing inks came from painters, and the first printing presses were designed after the smaller cloth and wine presses.

Gutenberg's press was designed from a fifteenth century wine press. The press was not the essence of Gutenberg's invention but merely a necessary accessory. Gutenberg also invented a unique printing ink for his process. Special ink was also crucial to Alois Senefelder's invention of lithography in 1798.

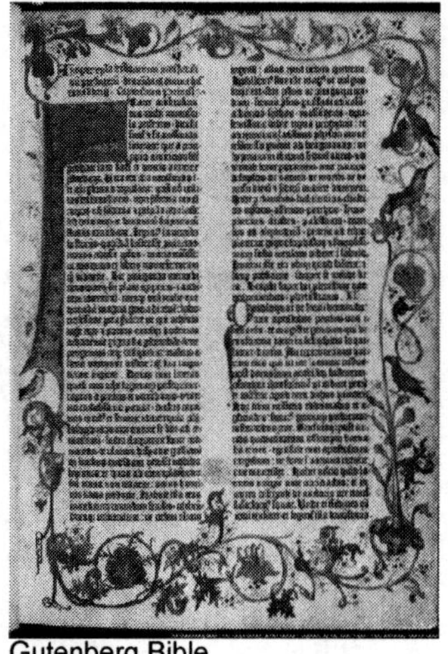

Gutenberg Bible

Gutenberg's Press
(History of the World)

Lithography, which means "stone printing," comes from the use of the smoothed surface of a porous Bavarian limestone upon which images for printing were drawn, engraved, or transferred. Lithography is a process that lends itself more to printing the smooth tones of pictures than it does printing from the raised metal surfaces of Gutenberg's process, which came to be called letterpress printing.

The First Reproduction of Images

Reproduction of images is believed to have first appeared between 4500 B.C. and 3500 B.C. Prior to the invention of writing, "stamp" seals made from stone or metal were carved in relief and used to indent ownership marks into moist clay. About 3500 B.C., cylinder seals containing duplicate relief, or raised, symbols introduced what eventually became the principle of modern rotary presses.

In 105 A.D. Ts'ai Lun discovered how to make "paper" from a pulp mixture of mulberry bark, hemp, linen rags, and water spread on a mat and sun-dried.

The Muslims carried the art of papermaking to Europe in 751. About 20 centuries elapsed between the invention of paper in China and its manufacture primarily from wood fiber. Movable relief images for "inkless printing" are believed to have been used in southern Asia Minor at about 1700 B.C. This involved the use of type-like relief symbols impressed into soft clay. The circular pattern of syllables found on the "Phaestos disk" on the island of Crete suggests an alphabetic structure and the first known use of re-usable images similar to relief type of Gutenberg's variety.

Papyrus and Paper: The Middle East, the Roman Empire, and Beyond

The history of communicating with the written word began between 1085 B.C. and 950 B.C. with the manufacture of papyrus and the use of pictography, or "picture words." While the ancient Egyptians had many uses for papyrus, its real importance was in the manufacture of what was then considered "paper." Papyrus grew along the banks of the Nile River and became the primary means of conveying pictographs and written words for centuries. In making paper from papyrus, the fibrous cores of papyrus stems were first cut into strips. The strips were then pounded, laid crosswise into a single sheet, pressed together, and dried. The sheet was then burnished with a stone to create a smooth surface, and a reed pen was used to write or draw on its surface.

The need for paper has been synonymous with the growth of printing. Paper is also the most expensive disposable commodity used in printing. Today there are moves to discover substitute materials for papermaking from pulp to overcome raw material shortages while continuing to provide paper having the strength, longevity, and surface characteristics needed for good printing.

Although paper had existed for nearly 2,000 years, it was Gutenberg's invention of movable type that spawned a revolution in papermaking and printing. In the 1450s, around the time that Gutenberg printed his Bible, only a few libraries in the West had collections of over 500 books. By 1500 more than 20 million books had been printed. The need for more paper on which to print all this information eventually yielded further innovations in papermaking. The basic process, however, has remained the same as that invented by Ts'ai Lun whereby a dilute mixture or slurry of fibers and water flow across a porous screen. The wet mass was then removed from the screen, the water pressed out, the sheet dried, and the end result was paper.

While traditional paper is still made from trees today, cotton and other seed fibers like flax are also used.

The manufacture of paper was improved with the invention of the Hollander beater—a device for beating rags to a pulp—around 1700. At the turn of the century Louis Robert, a French paper mill clerk, designed a paper-manufacturing machine. The work was continued in England, where engineer Bryan Donkin developed a commercial model during the period 1803–1812. The project was financed by Henry and Sealy Fourdrinier, prominent paper manufacturers, and the machine has been called the Fourdrinier ever since. While the Fourdrinier machine has been extensively improved over the years, it still remains the same in principle and is the dominant papermaking machine in use today. Fourdrinier machines are as long as several hundred yards and can make paper at 3,000 feet per minute.

The preservation of paper seems to get stronger with every suggestion of a "paperless" society. The promise of the twenty-first century is not of a paperless society but of one that embraces the quality of paper in a new and creative way. The development of synthetic paper that does not require pulp is well advanced. However, the promise of "digital paper"—paper that can display digital text—is a more recent development. Once the text is no longer needed, e.g., after reading, a page is electronically re-imaged with new text. This concept was developed by two organizations, the MIT Media Lab and the Xerox Palo Alto Research Center (PARC). This technology will be addressed later in this book.

Mixing pulp. (IPM)

Printing Emerges in the Far East

Long before Gutenberg, movable type for printing was invented in the Orient. A Chinese alchemist, Pi Sheng, invented type made from baked clay about 1043. A font of 60,000 wooden types was produced for the Korean ruler Wang Chen in 1313. Success of these type fonts was limited because of the large number and complexity of pieces required for printing Asian languages. Hence, there is evidence of metal type being used in the Orient by the fourteenth century and of the establishment of a central department of books.

Woodblock printing was also first used in the Orient. While historians credit the Chinese with woodblock printing on paper at least as early as 400 A.D., the Japanese Empress Shotoku commissioned the first "mass publication," a million copies of woodblock-printed sheets containing Buddhist prayers, between 764 A.D. and 777 A.D. Called the "million Buddhist charms," it is thought to be the oldest surviving example of woodblock printing in Japan and the earliest known specimen of printing. The Chinese produced the first block-printed book, *The Diamond Sutra*, in 868 A.D.

It is widely believed that the Buddhists must have introduced printing into Korea. The Korean king Ta-jong (1401–1419) was the first to carry out the idea of movable type made of copper. By 1403, forty-seven years before Gutenberg's first printing from movable type was known in Europe, type was also being cast of bronze.

The first book printed in Japan with movable type occurred in 1596. Because of the complexity of Asian alphabets, printing from woodblocks was much more practical and less expensive than printing from movable type.

Printing During the Renaissance

During the Dark Ages, rulers kept knowledge for themselves. Gradually, though, an interest in art, history, and science arose among the general populace, and the Renaissance—the rebirth of learning—emerged. Professional manuscript writing became an industry during this era. Then, in the midst of the Renaissance, something happened that spelled doom for the manuscript industry. It startled the world and propelled the rebirth of learning throughout Europe and beyond—It was the invention of printing in Europe spearheaded by Johann Gutenberg. Printing became an important aid for dissemination of knowledge; it induced persuasion or belief and became an essential tool for informing the masses.

Erasmus, in 1516, who was responsible for the first printing of the New Testament, was dedicated to using the emerging print technology of the time as a way of extending his mind through printed words. One of his chief interests was the publication of works by the famous Venetian printer and pioneer of finely printed popular editions, Aldus Manutius.

While printing's initial influence was on the humanities, the technology of printing eventually shaped the sciences as much as the arts. Modern sciences depend on information conveyed by exactly repeatable visual or pictorial statements.

Gutenberg, Movable Type, and the Printing Press

The art of typography is also attributed to Gutenberg, who designed his types to simulate the hand-lettered text of his time. Movable type worked perfectly for Gutenberg, for he had only the twenty-three basic letters of the Roman alphabet to deal with. He did more than just recycle the Chinese idea; he invented a mold that could turn out metal types that were exactly the same height. Thus, when the type was assembled, inked, and pressed against paper, the result was clean, uniform, and highly readable printing. Gutenberg could produce a hundred copies of a page in the time it took a scribe to make one or two pages.

Other printers in Europe were also experimenting with printing processes but Gutenberg was the only printer known to be using movable metal type. Essentially, there was a "race" to become the first to develop a workable method for producing and reproducing movable type as a logical extension of manuscript writing and to serve a growing demand for information Europe. Gutenberg's method of creating type by pouring molten lead into precise molds seemed to work best.

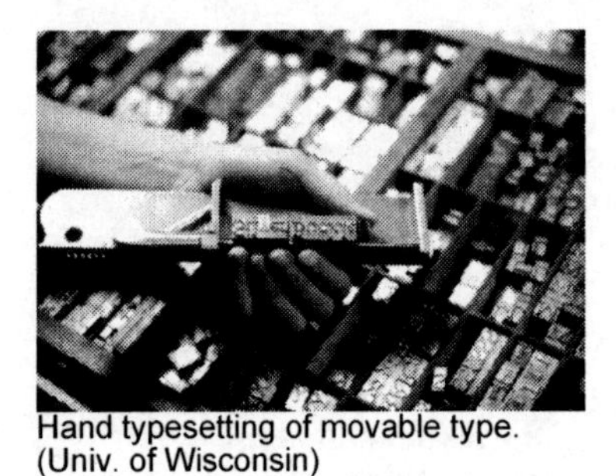
Hand typesetting of movable type. (Univ. of Wisconsin)

In 1454 Gutenberg and his assistant, Peter Schoeffer, began printing the Bible. Gutenberg printed approximately two hundred copies of what has come to be known as the "forty-two line Bible."

In the sixteenth century Peter Ramus introduced the concept of print-oriented classrooms in which books were available to all students. Before printing, much of the time in school was spent creating texts; the classroom tended to be a "scriptorium" with the student as editor-publisher. Printing changed learning. The book was the first "teaching machine" and also the first mass-produced commodity.

The proliferation of print put scribes out of work. Their plight resembles the plight of linotype operators in the 1960s with the demise of linotype machines. Numerous jobs were eliminated, mostly by attrition, when phototypesetting once again revolutionized the printing industry. However, many linotype operators were able to retrain as paste-up and mechanical "artists," while others found positions in sales and service.

After Gutenberg's invention, for the first time hundreds of readers owned identical copies of the same book, and the book one read in one city was the same someone else was reading in another city. Gutenberg's invention generated the first "information overload," and today's digital technology is threatening us with another information overload. There are several examples of how the impact of Gutenberg's invention equates with World Wide Web developments.

Gutenberg's Influence Abounds

At about the same time Gutenberg was perfecting his invention for relief or letterpress printing, a process employing the opposite principle of printing from an engraved or depressed surface was being developed. Called "intaglio," this process was used for copper engraving as early as 1446. Metal smiths produced intricate designs by carefully inscribing or scratching depressions into a soft copper surface. Ink placed on the finished engraving and wiped off the top surface would remain in the recessed depressions; this ink could then be transferred to paper pressed against the image. Today's steel engraving and rotogravure processes use the intaglio principle.

In 1470 Nicholas Jenson, a famous Italian printer and typographer, perfected the clarity, beauty, and utility of the Roman typeface—a monumental contribution to the history of the graphic arts. Jenson's first Roman typeface introduced thick and thin strokes and serifs. His alphabets—based on simplicity and appealing stroke and space proportions—still influence type designed more than five hundred years later. William Caxton brought printing to England in 1476, one year after he had printed the first book in English in the Netherlands.

One of the greatest figures in publishing history, Aldus Manutius established a printing company in Venice during 1495. Although his works are distinguished by exceptional typographical beauty, Manutius wished to make it possible for common people to afford editions of Greek and Latin classics. He published "pocket-sized" works of Aesop, Aristotle, Euripedes, Homer, Plato, Plutarch, and others at a low cost. In 1501, he printed the first book containing italics.

Printing, Reading, and Consumer Markets

Typesetting altered language from a means of perception and exploration to a portable commodity. Printed books themselves were the first uniform, repeatable, and mass-produced items in the world. It was not until the nineteenth century that production-to-consumption concepts for print were fully realized. It took two centuries to understand the potential of the printing industry to mass-produce printed products as commodities.

The printing industry has experienced similar "concept-to-market" success delays, though of shorter time spans, with the introduction of nearly every new technology. The linotype machine was introduced in 1885 but was not considered a commercial success until 1905. Electronic color separation scanners were available in the 1940s but did not become commercially viable until the 1970s. It took the Gannett Corporation more than ten years to realize its first profits using satellite transmission to produce *USA Today* in color at multiple sites simultaneously. The concept and technology of on-demand digital printing has created enormous market anxiety for those who have invested in these technologies and are attempting to develop niches for them. These newer technologies are covered in detail later in this book.

Over the centuries, the development of multiple print technologies created new challenges. Books flourished as a result of the development of copper engravings. There was also the choice between the woodcut—which could be inserted in the same print form as the type characters and print on the same press—and the copperplate engraving—which was considered superior as early as the end of the sixteenth century but which required a special, more costly, printing press. The use of copperplates required that the printer treat text and illustration as distinct elements and print them separately. Broadsheets and flysheets provided the opportunity to engrave the text onto copper plates along with the picture.

Lithography Solves Problems

Senefelder's invention of lithography in 1789 solved some of the problems related to printing from raised metal type, woodcuts, and copperplate engravings. Not only was lithography a process that simplified the reproduction of multi-color pictures and improved tonal gradations, but it also made it easier to combine pictures and type on one plate or image carrier. Additionally, lithography provided an efficient means of printing "alphabets" made of thousands of complex symbols such as those used in the Far East.

Lithographic stones were made from porous Bavarian limestone upon which images were drawn, engraved, or transferred through a chemical process that separated the image and non-image areas. It is based on the principle that "grease and water do not mix." Besides inventing special "greasy" ink for the process—which was used to write or draw on the stone in reverse—Senefelder also devised a method for transferring the images to a special paper which he called a "transfer sheet." Hence, he was able to draw or write a "right-reading" image that was then transferred, or "offset," as a "wrong-reading" (or reverse) image on the stone. The image would then be transferred again as "right-reading" onto the paper.

To make prints from the image, Senefelder sponged the stone with a gum arabic solution and rolled on greasy ink that would be accepted by the greasy image lines but not by the dampened non-image areas. He then pressed a sheet of paper over the surface to print the image.

Lithography began to flourish as a medium for artistic expression. Because it was the most economical way to reproduce illustrations, lithography became the most popular. Sometimes color was applied by hand to fine art lithographic reproductions. In 1825, for example, the Spanish painter Goya used such techniques to produce his famous "Bullfighter" series of lithographs. French artists later made lithographs in many colors and with tone gradations. In the United States, businessman Nathaniel Currier and artist J. Merrit Ives became partners in the production of the noted Currier & Ives lithographs. Toulouse-Lautrec and Whistler were also among the many well-known artists who used lithography as their process of choice in producing art prints.

Lithography eventually became the predominant process for nearly all printing applications. Lithography still dominates today, albeit with highly sophisticated printing presses using thin printing plates such as aluminum and other flexible metal image-carriers that replaced stones in the 1930s and 1940s.

Printer Ira Rubel replaced the transfer sheet with rubber "offset blankets" in the early 1900s, and the lithographic process became known as "offset lithography."

In the early 1900s twelve offset lithographic presses were built by the Potter Printing Press Company, which merged in 1927 with the Harris Company to form Harris-Seybold-Potter Co. The Harris Corporation went on to become one of the largest manufacturers of highly sophisticated printing presses of the twentieth century and experienced further growth when acquired in the mid-1990s by the Heidelberg Corporation, the world's largest producer of printing presses. Today the original Harris Corporation is owned by Goss International.

Even with developments in commercial offset lithographic presses, stone lithography still has a place as an art form for original printmaking.

Printing Comes to North America

The first printer in North America was Juan Pablos. He arrived in 1539 and set up his press in Mexico City. Other printers followed, and Mexico became a printing center in North America. It would be a hundred years, though, before printing reached the "New World," or what is now the United States.

The first known printing in the United States was a one-page oath produced around 1638 in Massachusetts. The first book printed in Colonial America was a book of psalms printed by Matthew Day in 1640. Later came prayer books, almanacs, Bibles, sermons, lesson books, and newspapers—all produced on wooden presses.

Steven Daye established a press in Cambridge near Boston. His "Freeman's Oath," produced on a single sheet, is believed to be the first printing in the American colonies. As printing grew in New England, so did the need for a ready supply of paper to feed the presses. William Rittenhouse started the first paper mill in 1690 near a small river outside Philadelphia.

Early Newspapers in America

The Boston News-Letter, published in 1704, is considered the first American newspaper, although Benjamin Harris had printed a single issue of a one-page *Publick Occurrences* in 1690. Established by John Campbell and printed in the shop of Bartholomew Green, the *News-Letter* was published until 1776.

Benjamin Franklin was a skilled scientist, author, editor, publisher, and printer. His *Pennsylvania Gazette*, started in Philadelphia in 1732, was a well-respected newspaper. Among Franklin's many writings was the series of *Poor Richard's Almanacs*, published annually between 1732 and 1758 in runs as large as 10,000 copies on a hand press.

In 1814, after centuries of using history's oldest printing machine, the hand press was finally superseded and virtually discontinued forever as a production tool when Friederich Koenig sold his first power-driven press to *The London Times*. Koenig's invention immediately lowered printing and composing costs by 25 percent and made possible the production of inexpensive and longer publications of every kind.

In 1822 Peter Smith, an American associated with the R. Hoe & Co., devised a machine that was in many ways superior to any concept up to that time. Its frame was of cast iron, and in place of a screw with a lever he substituted a toggle joint, which was simple and effective. In 1827 an invention by Samuel Rust was a great improvement on the Smith press. Instead of being all cast iron, the press frame's uprights at the side were hollowed for the admission of wrought-iron bars, which were securely riveted at the top and bottom of the casting. This gave additional strength and greatly diminished the amount of metal used in construction. The Rust patent was purchased by R. Hoe & Co. and improved. The new invention was known as the "Washington Hand Press," and in principle and construction has never been surpassed by any hand printing press. The Washington Press was manufactured in great numbers and sold around the world—only to be superseded by the universal adoption of the cylinder press. More than six thousand Washington Presses were made and sold by R. Hoe & Co.

Harper's Publications

From the mid-1800s to the mid-1900s, a series of concurrent developments greatly widened public access to print. Steam-powered cylindrical presses, stereotype plates, and paper mass-produced from wood pulp greatly increased the output and decreased the price of printed material. By the end of the nineteenth century, several hundred times as many pages were printed per capita as at the beginning of the century. The invention of the telegraph and the laying of the Atlantic cable enormously enhanced the speed and efficiency with which the press could convey news. Additionally, the completion of a rail network made practical the swift nationwide distribution of magazines and books. Nearly universal elementary education and widespread literacy opened a broad public market for print. With the "penny newspaper" and inexpensive magazines and books, print became a mass medium.

Mass production of printed material became easier in the nineteenth century through a series of improvements in press designs. The first power-driven cylinder press sold to *The London Times* in 1814 speeded printing by carrying paper in a circular fashion around a cylinder to contact a flat form of type below. In 1846 R. Hoe & Co. produced a type-revolving press—the first to use a rotary, cylinder-to-cylinder, principle. Its design required that forms containing loose pieces of type be wrapped around one cylinder. This design was replaced in 1871 with a rotary press that used single-piece stereotype plates. George P. Gordon devised a treadle-operated version of a platen press in 1850. In 1856 William Bullock produced the first web-fed, perfecting press, capable of printing on both sides of a continuous roll of paper.

Charles Mahon, third earl of Stanhope, re-established stereotype printing in England and perfected his method of stereotyping from plaster molds. His process offered obvious advantages wherever books were being frequently reprinted such as bibles, prayer books, and schoolbooks. Stereotypes greatly reduced the wear of type, and a stereotype plate would not break into pieces as a form of type often did.

The first line of Gordon job presses was put on the American market by George Phineas Gordon in 1851 and was used for efficient printing production well into the twentieth century.

The Linotype Machine

The Industrial Revolution contributed one of its most significant technologies when on July 3, 1886, Ottmar Megenthaler, a German immigrant, demonstrated to the *New York Tribune* a machine that today is regarded as one of the ten greatest inventions of all time: the linotype machine. Type for nearly every newspaper in the world and most books in the United States would eventually be composed on Mergenthaler's machine or on machines that achieve similar results such as the Harris Intertype or the Monotype machine. When keys on a linotype are struck, pieces of brass punched with characters are brought together into a line, automatically spaced, and moved to a mold where molten metal is injected into the indented characters to produce slugs of metal type in complete lines. The Harris Intertype operated on the same principles, whereas the Monotype machine created type from individually cast characters automatically formed into lines. The Monotype had the added distinction of being driven by punched paper tape produced on a keyboard that was separate from the typesetter or caster. Punched paper tape was used in this capacity for decades before the same concept was applied to computers.

Linotype Machine
(Greenfield Print Shop)

The linotype and similar technologies completely revolutionized printing and publishing, allowing far more pages to be printed in newspapers and magazines than was possible with handset movable type. Today, the publishing of books, newspapers, and other printed items depends on "keyboarding," albeit through computers. The linotype machine and related technologies were completely obsolete by the mid-1960s.

The first practical typesetter that did not require molten hot metal to produce images was the Fotosetter tested by the Harris Intertype Corporation in 1946. While this machine's outward appearance resembled the hot metal linecasters of the day, its imaging system was quite different. The small brass matrices (mats) that circulated inside the machine contained not tiny letter molds but film negatives, each with an image of a letter, numeral, or symbol. As the matrix passed a specified point in its travel through the machine, a beam of light flashed through the negative in its side, exposing the image onto photographic paper or film. The Fotosetter was soon displaced by more advanced and faster photo-optic systems of simpler design. Hence, a synergy of photography and typesetting became the mainstay of typesetting technology until the mid-1980s, when desktop publishing took over the typesetting function for most printing applications.

Typography

With the advent of printing technology came the need to develop type designs that reproduced properly on a printing press within the limitation that press technology imposed. For example, the image produced by squeezing ink on paper was different from an image produced by writing on paper. Thus the art of typography was developed and, along with it, the profession of "typographer." Typography relates to the aesthetics of type and type design, while typesetting refers to the technical operation of composing, setting, or arranging type for printing.

Thousands of type styles exist today and some have evolved from the work of a few individuals of the eighteenth century. Three great names in early type design are William Caslon and John Baskerville of England and Giambattista Bodoni of Italy. Each developed widely acclaimed type designs that bear their last names and became the forerunners of major divisions of Roman type used for printing. A more contemporary twentieth century typographer was Frederick W. Goudy. During his lifetime Goudy designed more than one hundred typefaces, considered by some experts to be the most beautiful in existence.

Wood Type (Impressions)

Printing in the Electronic Era

Lasers, microwave relays, cable communication, satellite transmission, the application of microchips and microprocessors, and related technologies are all developments of the twentieth century. These technologies were part of a synergy that led to specific products for printing, including electronic color separation scanners, imagesetters, front-end and integrated systems, and digital presses. The combined influence of these technologies and products became part of two twentieth century revolutions in the printing industry: audiovisual and multi-media. Developments in these areas provided a range of alternatives to communication typically carried by print and a means of communication not achievable through print.

Computers and Desktop Publishing: Everybody's Business

Innovations in integrated circuits, transistors, and microprocessors using silicon chips have dramatically influenced the size, speed, and capacity of computers. These innovations have improved the printing industry in two significant stages. The first is that nearly every piece of printing equipment that has been developed since the mid-1980s is driven by a microprocessor. The second is that the miniaturization of and cost reductions in microprocessor-driven technology have given nearly any person access to printing technology. This is most obvious in desktop publishing technology, where image generation tasks can be performed for under $1,000 that cost hundreds of thousands of dollars a decade earlier. Through such accessibility, the author of print can now also be the producer of print.

The second stage of the "author as producer" trend is the rapidly developing availability of high-quality, high-speed printing systems for mass production of text and pictorial images produced on the "desktop." The roots of this second stage were planted in 1938 when Chesley Carlson invented a process that helped make possible the office copier, a "printing plant" in the office. Carlson's invention became known as xerography, or "dry writing," and was accomplished by reflecting the image of an original document from a mirror onto an electrically charged selenium drum. A copy was created as special dry-ink powder (toner), attracted to the dark image on the drum, was transferred and then fused to paper. Carlson's invention was an anomaly for printing systems of his time but represented a paradigm for printing technology concepts for the twenty-first century. Sometimes referred to as "electrophotography," Carlson's xerography has become increasingly important in "on-demand" printing where electronic printing and sophisticated bindery and finishing attachments have made such systems competitive with traditional lithographic presses.

How long have concepts of desktop publishing been around? One can trace the origins of desktop publishing back to the days of medieval scribes and students of that time, who had to be paleographer, editor, and publisher of the authors they read. A manuscript book was costly; the simplest way of obtaining books was for the teacher to dictate the texts to the students, i.e., "desktop publishing." Such a commercial venture on the part of students who wrote and sold books assured teachers large audiences and, in some cases, substantial revenue.

Through the first half of the twentieth century, the competing technologies of radio and television expanded methods of communications. However, printing continued to grow. Electronics brought to printing capabilities that enhanced quality and the speed in which printing could be produced and distributed. In some cases, print moved to computer screens, but books, magazines, newspapers, manuals, and packaging all continued to grow in demand. In fact, desktop publishing has made printing more affordable and more available. Using computers, today's printer can create complex layouts, change colors, and manipulate images in a matter of minutes.

Today desktop publishing is everybody's business. From home user to the classroom to the business world, desktop publishing has become a necessary way of preparing and distributing printed information.

The modern desktop publishing era began in the mid-1980s with the development of Aldus PageMaker, which then became Adobe PageMaker. Combined with the introduction of the Apple LaserWriter, a PostScript printer, and PageMaker for the Macintosh computer, this rounded out the early technology of desktop publishing, a term "coined" by Aldus Corporation founder Paul Brainerd in 1985. The introduction of PostScript by Adobe was key to all of this. PostScript is a page description language that allows computers to communicate with printers. In 1985 Apple Computer Company produced the LaserWriter, the first laser printer to use PostScript.

Early Apple Macintosh desktop publishing system. (Apple Computer)

With these early developments, it became evident that desktop publishing hardware and software were going to revolutionize how printing took place, and there was a need to expand desktop publishing capabilities beyond the Macintosh computer platform. While Macintosh became the platform of choice in the graphic arts industry, PCs were the platform of choice in the general business world. Therefore, in 1987 PageMaker for the Windows PC platform was introduced, and in 1990 the Microsoft Corporation made available Windows 3.0, which gave PC users many of the desktop publishing abilities previously only available on the Macintosh platform.

By 2003, Hewlett-Packard LaserJets and Apple LaserWriters could still be purchased. However, there were also numerous other printers and printer manufacturers to choose from as well. By the turn of the century, PostScript and PageMaker were well advanced though upgrades and improvements, and are now used regularly not only in the graphic arts industry but by businesses of all sorts.

PageMaker and QuarkXPress eventually would become the main desktop publishing applications. However, more recently Adobe's InDesign has competed favorably with QuarkXPress and is now commonly used on PC and Macintosh platforms.

While the Macintosh remains the platform of choice for most professional desktop publishing, many other forms of desktop publishing hardware and software became available in the 1990s primarily for PC Windows users.

The main applications for Macintosh and PC platforms are described in detail later in this book.

The advent of desktop publishing has broadened the scope of the graphic arts to include not only traditional printing establishments, but publishers, designers, advertising agencies, print buyers, service bureaus, print brokers, and basically any individual who desires to produce something as simple as a one-color letterhead or a complex multicolor brochure. Desktop publishing has allowed the author and originator of an idea to become the producer of print media.

3

History of the Internet and World Wide Web

The Concept of the Internet

The concept of the Internet has been around since the early 1960s. In the late 1980s, the creation of the World Wide Web (WWW) addressed the great desire for those using the Internet to receive and send communications in color and pictures rather than text alone. Hence, the Internet came first and the WWW was a product of it.

The Internet is a network of interconnected computers. It is a community of people. It is the "global village" that Marshall McLuhan considered television to be in the 1960s. However, the Internet surpasses television as a "global village" because it allows interaction. If you do not know who Marshall McLuhan was, plug his name into any search engine and enjoy the ride. It'll be worth it.

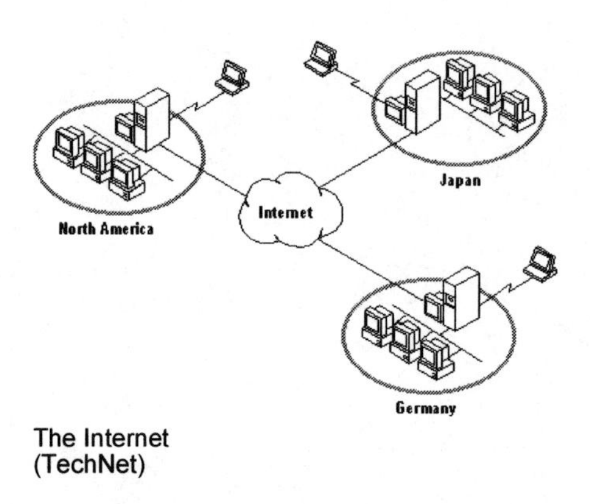

The Internet
(TechNet)

Visionary Thinkers

The Internet was the result of some visionary thinkers in the early 1960s who saw great potential value in allowing computers to share information on research and development in scientific and military fields. J.C.R. Licklider of MIT first proposed a global network of computers in 1962 and moved over to the Defense Advanced Research Projects Agency (DARPA) to head development of it. Leonard Kleinrock of MIT and later UCLA developed the theory of packet switching, which would form the basis of Internet connections. Lawrence Roberts of MIT connected a Massachusetts computer with a California computer in 1965 over telephone lines. This showed the feasibility of wide area networking, but also that the telephone line's circuit switching was inadequate. Kleinrock's packet switching theory was confirmed. Roberts moved over to

DARPA in 1966 and developed his plan for ARPANET (Advanced Research Projects Agency).

The ARPANET

The Internet, then known as ARPANET, was brought online in 1969 under a contract led by the renamed Advanced Research Projects Agency (ARPA) that initially connected four major computers at universities in the southwestern United States (UCLA, Stanford Research Institute, University of California Santa Barbara, and the University of Utah).

The four computers then grew with other computers to form the Internet. Today online "traffic" sometimes slows down because people are using bigger chunks of data (pictures, audio, and video). This situation is rapidly improving with added computing power and speed enhancements.

So, the Internet was initially a project called ARPANET funded by the Department of Defense that was used to communicate when certain defense nodes went down. The MILNET, an independent defense and intelligence information service, was developed in the mid-1970s to expand ARPANET.

The World Wide Web (WWW)

Swiss academician Tim Berners-Lee developed the World Wide Web (WWW), an invisible hypermedia system that links text, voice, and video files by clicking on an icon. The WWW has spawned the increased use of the Internet. E-mail continues to be the most popular feature, though Web browsing and e-commerce has grown rapidly. The Web is also used strictly for entertainment.

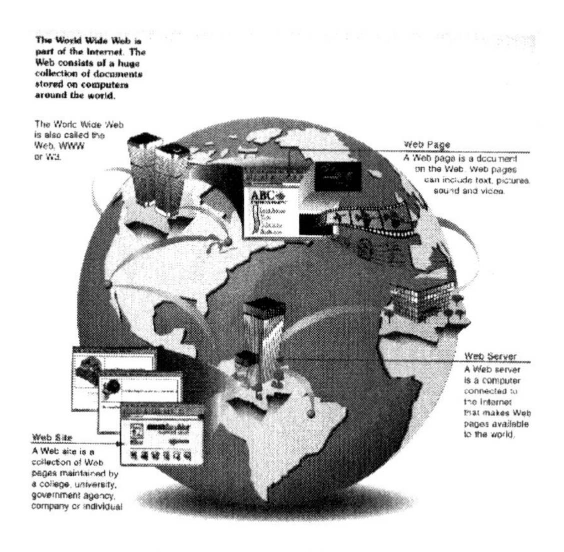

The global reach of the World Wide Web.
(Univ. of Minnesota)

The invention of the telegraph, telephone, radio, and computer set the stage, and the WWW has revolutionized communications. The Internet broadcasts to the world and as such serves as a mechanism for information dissemination and interaction between individuals in any geographic location.

The original ARPANET grew into today's Internet and was based on the idea that there would be multiple independent networks beginning with the ARPANET as the pioneering packet switching network. It soon grew to include packet satellite networks, ground-based packet radio networks, and other networks.

The Defense Advanced Research Projects Agency (DARPA)

The Defense Advanced Research Projects Agency (DARPA) became involved in 1972, when open-architecture networking was introduced for the ARPANET. Open-architecture means that the standards and design are open to the public and the public may use add-ons or modifications. DARPA developed a protocol that would eventually be called the Transmission Control Protocol/Internet Protocol (TCP/IP); this new protocol allowed external agencies to connect to the ARPANET.

TCP/IP led to the development of the Ethernet at the Xerox Palo Alto Research Center (PARC) and then to the establishment of Local Area Networks (LAN) in the late 1970s.

An initial motivation for the Internet was resource sharing that would allow users to access time-sharing systems attached to the ARPANET. Electronic mail has probably had the most significant impact of the innovations from that era. E-mail provided a new model of how people could communicate with each other and changed the nature of collaboration—first in the building of the Internet itself and later for much of society.

In 1973 DARPA initiated a research program to investigate techniques and technologies for interlinking networks. The objective was to develop communication protocols that would allow networked computers to communicate transparently across multiple linked networks. This was called the "Internetting Project," and the system of networks that emerged would come to be known as the Internet.

Transmission Control Protocol (TCP) and Internet Protocol (IP)

Transmission Control Protocol (TCP) and Internet Protocol (IP) were essential in establishing open-architecture connectivity.

The Ethernet is a LAN developed by Bob Metcalfe at the PARC in 1973 in cooperation with DEC and Intel. It is now the dominant network technology in the Internet, and PCs and workstations are the dominant computers.

A major shift occurred as a result of the increase in scale of the Internet. To make it easy for people to use the network, hosts were assigned names so that it was not necessary to remember a set of numeric addresses. The Domain Name System (DNS) was invented by Paul Mockapetris of the University of Southern California. The DNS permitted a mechanism for resolving host names—such as www.acm.org—into an Internet address.

The early implementations of TCP were done for large time-sharing systems. When desktop computers first appeared, TCP was considered too large and complex to run on a personal computer. David Clark and his research group at MIT set out to show that a compact and simple implementation of TCP was possible and produced one—first for the Xerox Alto (the early personal workstation developed at Xerox PARC) and then for the IBM PC. They showed that workstations, as well as large time-sharing systems, could be a part of the Internet.

Computer experts, engineers, scientists, and librarians were the primarily users of the early Internet. There was nothing friendly about it. There were no home or office personal computers in those days, and anyone who used the Internet—whether a computer professional or an engineer or scientist or librarian—had to learn to use a very complex system.

E-mail was adapted for ARPANET by Ray Tomlinson of BBN Technologies in 1972. BBN (Bolt, Beranek & Newman) became part of Bell Systems Verizon Communications Inc. He picked the @ symbol from the available symbols on his teletype to link the username and address. The telnet protocol, enabling logging onto a remote computer, was published as a Request for Comments (RFC) in 1972. RFCs are a means of sharing developmental work throughout a community. The ftp (file transfer protocol), which enables file transfers between Internet sites, was published as an RFC in 1973. From then on RFCs were available electronically to anyone with access to the ftp protocol.

Libraries began automating and networking their catalogs in the late 1960s independent from ARPANET. Automated catalogs, which were not very user-friendly at first, became available to the world, first through telnet and many years later through the Web.

The Internet matured in the 1970s as a result of the TCP/IP architecture; it was adopted by the Defense Department in 1980 and universally by 1983.

Newsgroups followed. These discussion groups focusing on a topic provided a means of exchanging information throughout the world. Many Internet sites took advantage of the availability of newsgroups. It was a significant part of the community building that took place on the networks.

Both public domain and commercial implementations of the roughly 100 protocols of TCP/IP protocol suite became available in the 1980s. Widespread development of LANS, PCs and workstations in the 1980s allowed the Internet to flourish.

The Internet is as much a collection of communities as a collection of technologies, and its success is largely attributable to satisfying basic community needs as well as to allowing the community to forward the technology and its capabilities effectively. This community spirit has a long history beginning with the early ARPANET, where researchers worked as a close-knit community to accomplish the initial demonstrations of packet switching technology. Starting in the early 1980s, the Internet grew beyond its primary research roots to include a broad user community and increased commercial activity. Increased attention was paid to making the process open and fair.

Commercialization of the Internet involved the development of competitive, private network services as well as commercial products implementing Internet technology. In the early 1980s, dozens of vendors were incorporating TCP/IP into their products because they knew there were buyers. Unfortunately they lacked real information about how the technology was supposed to work and how the customers planned to use the Internet for networking.

CERN

The Ethernet made its appearance at the European Laboratory for Particle Physics—known as CERN—in 1983 when an initial stretch of Ethernet cable arrived to support a demonstration of TCP/IP protocols. TCP/IP had been adopted as a defense standard three years earlier in 1980, and by 1983 ARPANET was being used by a significant number of defense research and development and operational organizations. By 1985 the Internet was well established as a technology supporting a broad community of researchers and developers and was beginning to be used for daily communication.

In 1986 the U.S. National Science Foundation (NSF) initiated the development of the NSFNET, which provides the major backbone communication service for the Internet. The National Aeronautics and Space Administration (NASA) and the U.S. Department of Energy contributed additional backbone facilities. The following year, the European Internet occurred when major international backbones provided connectivity to computers on a large number of networks.

Various consortium networks provide “regional” support for the Internet, and "local" support is provided through research and educational institutions. Within the United States, much of this support has come from the federal and state governments, but a considerable contribution has been made by industry. In Europe and elsewhere, support arises from cooperative international efforts and through national research organizations.

Between 1985 and 1988, the coordinated introduction of TCP/IP within CERN made excellent progress. The Cray computer represented CERN's first "supercomputer" according to U.S. military and commercial standards, and a serious security system was erected around it. By 1989 CERN's Internet facility was ready to become the medium within which Tim Berners-Lee would create the World Wide Web.

During the late 1980s, the population of Internet users and network constituents expanded internationally and began to include commercial facilities. Indeed, the bulk of the system today is made up of private networking facilities in educational and research institutions, businesses, and government organizations around the globe.

BITNET (Because It's Time Network)

Similarly, BITNET (Because It's Time Network) connected IBM mainframes around the educational community and the world to provide mail services beginning in 1981. Listserv software was developed for this network and later others. Gateways were developed to connect BITNET with the Internet and allowed exchange of e-mail, particularly for e-mail discussion lists.

As the commands for e-mail, FTP, and telnet were standardized, it became a lot easier for non-technical people to learn to use them. It was not easy by today's standards by any means, but it did open up use of the Internet to many more people—in universities in particular. Departments other than the library, physics, and engineering found ways to make good use of the “nets”—to communicate with colleagues around the world and to share files and resources.

While the number of sites on the Internet was small, it was fairly easy to keep track of the resources of interest that were available. But as more universities—and their libraries—connected, the Internet became harder to track. There was a need for tools to index the available resources.

The first effort, other than library catalogs, to index the Internet was created in 1989, as Peter Deutsch and his team at McGill University in Montreal created an archiving tool for ftp sites, which they named Archie. This software would periodically reach out to all known openly available ftp sites, list their files, and build a searchable index of the software. The commands to search Archie were

AT&T's UNIX commands, and it took some knowledge of UNIX to use it to its full capability.

During the early 1990s, Open Systems Interconnection (OSI) protocol implementations also became available and, by the end of 1991, the Internet had grown to include approximately 5,000 networks in over three dozen countries, serving over 700,000 host computers used by more than 4,000,000 people.

Gopher

In 1991 the first user-friendly interface was developed at the University of Minnesota. The university sought to develop a simple menu system to access files and information on campus through their local network, and its demonstration system was called a Gopher. The Gopher proved to be prolific, and within a few years there were more than 10,000 Gophers around the world.

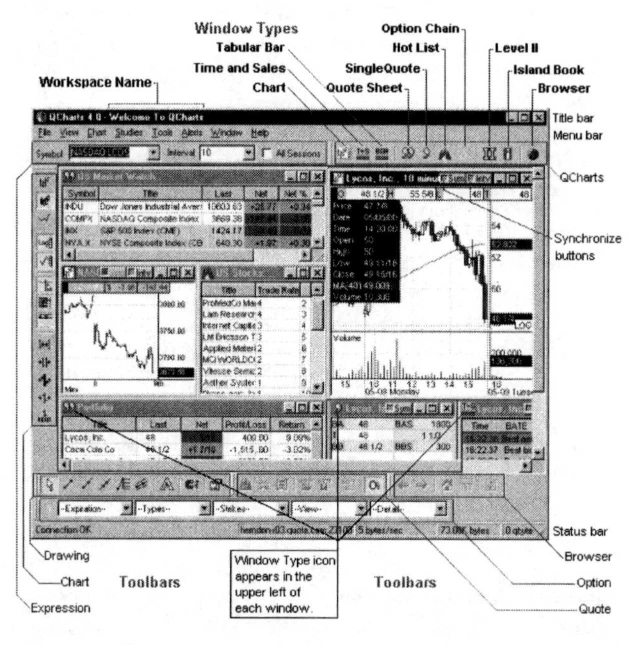

A typical Web page.
(Quote.com)

A Gopher system requires no knowledge of UNIX or computer architecture to use: a number is typed or clicked to select the desired menu selection. Gopher's usability was enhanced when the University of Nevada at Reno developed the VERONICA searchable index of Gopher menus. It was purported to be an acronym for Very Easy Rodent-Oriented Netwide Index to Computerized Archives. A "spider" crawled Gopher menus around the world, collecting links and retrieving them for the index. It was so popular that it was very difficult to connect to, even though a number of other VERONICA sites were developed to ease the load. Similar indexing software was developed for single sites and was called JUGHEAD (Jonzy's Universal Gopher Hierarchy Excavation and Display).

In 1989 another significant event took place that made the Internet easier to use. Tim Berners-Lee and others at CERN proposed a new protocol for information distribution. This protocol, which became the World Wide Web in 1991, was based on hypertext—a system of embedding links in text to link to other text.

Mosaic

The development in 1993 of the graphical browser Mosaic by Marc Andreessen and his team at the National Center for Supercomputing Applications (NCSA) gave the protocol its big boost. Later Andreessen would become the brains behind the Netscape Corp., which produced the most successful graphical type of browser and server until Microsoft developed Internet Explorer.

Because the Internet was initially funded by the government, it was limited to research, education, and government uses. Commercial use was prohibited unless it directly served the goals of research and education. This policy continued until the early 1990s, when independent commercial networks began to grow. It then became possible to route traffic across the country from one commercial site to another without passing through the government-funded NSFNet Internet backbone.

Delphi

Delphi was the first national commercial online service to offer Internet access to its subscribers. It opened up an e-mail connection in July 1992 and full Internet service in November 1992. All pretenses of limitations on commercial use disappeared in May 1995 when the National Science Foundation ended its sponsorship of the Internet backbone, and all traffic relied on commercial networks. AOL, Prodigy, and CompuServe came online.

Microsoft's full-scale entry into the browser, server, and Internet service provider market completed the major shift over to a commercially based Internet. The release of Windows 98 in June 1998 with the Microsoft browser well integrated into the "desktop" showed Microsoft's determination to capitalize on the enormous growth of the Internet.

The development of high-speed connections came next. The 56K modems were not fast enough to carry multimedia, such as sound and video, except in low quality. But new technologies many times faster—such as cable modems, digital subscriber lines (DSL), and satellite broadcast—are now widely available.

During this period of enormous growth, businesses entering the Internet arena scrambled to find economic models that work. Free services supported by advertising shifted some of the direct costs away from the consumer. Services such as Delphi offered free Web pages, chat rooms, and message boards for community building. Online sales have grown rapidly for such products as books, music CDs, and computers, but the profit margins are slim when price comparisons are easy, and public trust in online security is still questionable. Business models that have worked well are portal sites that try to provide everything for everybody including live auctions. AOL's acquisition of Time-Warner was the largest merger in history when it took place and shows the enormous growth of Internet business.

It is becoming increasingly clear that many free services will not survive. While users still expect free services, there are fewer providers who can provide it. The value of the Internet and the WWW is undeniable, but there is a lot of shaking out to do, and management of costs and expectations to contend with.

E-mail

E-mail has been significant in all areas of the Internet, and in the last few years has spawned a new phase of commercialization. Originally, commercial efforts consisted mainly of vendors providing basic networking products, and service providers offering connectivity and Internet services. The Internet has now become almost a "commodity" service, and much of the latest focus has been on the use of this global information infrastructure for support of other commercial services. Accelerating this trend has been the widespread and rapid adoption of browsers and World Wide Web technology that allow easy access to information linked throughout the globe. Products are available to facilitate the provisioning of that information, and recent technology has been aimed at providing increasingly sophisticated services on top of the basic data communications.

The Internet has changed much in the past few decades. It was conceived in the era of time-sharing but has survived into the era of personal computers, client-server and peer-to-peer computing, and the network computer. It was designed before LANs existed but has accommodated that new network technology, as well as the more recent ATM and frame switched services. Initially it was designed to support a range of functions from file sharing and remote login to resource sharing and collaboration, and it has spawned electronic mail and the World Wide Web. But most important, what started as the creation of a small band of dedicated researchers and has grown to be a billion-dollar industry.

The Internet has not finished changing. Although a network in name and geography, the Internet is a creature of the computer, not the telephone or television industry. It must continue to change and evolve if it is to remain relevant. It has already changed to provide such new services as real-time transport in order to support, for example, audio and video streams. The availability of pervasive networking along with powerful affordable computing in portable form—i.e., laptop computers, two-way pagers, PDAs, cellular phones—makes possible a new paradigm for communications.

This evolution will continue to bring new applications. Internet telephone and Internet television are two examples. It is evolving to permit more sophisticated forms of pricing and cost recovery, a requirement in the commercial world, and it is changing to accommodate yet another generation of underlying network technologies with different characteristics and requirements—from broadband residential access to satellites. New modes of access and forms of service will spawn new applications, which in turn will drive further evolution of the "net" itself.

The most pressing question for the future of the Internet, as well as the broader range of graphic communication, is not how the technology will change, but how the process of change and evolution itself will be managed.

4

Technological Transitions in Graphic Communication

The history of graphic communication shows that the development of new technologies accelerates at a faster rate with each year while the technology becomes more complex and versatile. Consider the following.

It was 430 years from the time that Gutenberg invented the process of duplicating movable type in 1456 to Mergenthaler's invention of the Linotype machine in 1886. However, it was only 68 years between the introduction of the Linotype and the invention of phototypesetting by the Harris Corporation in 1954. This technological transition time compression trend became evident in the graphic arts with the introduction of satellites to transmit data to multiple sites simultaneously in the production of *The Wall Street Journal* by Dow Jones in 1968, only 14 years later. Computer graphics on cathode ray tubes took hold six years later in 1974—first for engineering, architectural rendering, automobile design, and then for graphic arts applications. Three years later, in 1977, the Mitsushita Corporation in Japan created a prototype of a color (TV) monitor that printed out anything on its screen in full color. This was the forerunner of today's computer monitors, interfaced with printers, for the production of desktop color. What Mitsushita did was build an inkjet printing system into a television set to demonstrate the potential of integrated media. Two years later in 1979 came the first practical combination of multiple graphic arts technologies from Scitex in Israel that demonstrated how nearly all prepress functions (page layout, retouching, typesetting, graphic arts photography, and image assembly) can be incorporated in one integrated system. Called the Response 300, Scitex planted the seed that led to the demise of traditional prepress departments in the graphic arts. One year later in 1980 Gannett demonstrated how satellite transmission could be used to disseminate color data in the production of *USA Today*.

By 1980, technological transitions in graphic communication were occurring so rapidly that it became impossible to point to individual development to in illustrate the speed with which technology was changing. It became necessary to point to changes occurring in a group of years. For example, 1980 through 1985 resulted in the U.S. government's deregulation of telecommunication, which opened the airwaves for broad commercial applications of communication media and provided great promise for the growth of digital printing and the Internet. This was a macro-technological development. From 1985 through 1990 there was the birth, growth, and maturity of low-cost desktop publishing systems—a micro-technological development. The macro-technology of satellite transmission and the micro-technology of desktop publishing eventually combined. Multimedia applications flourished between 1990 and 1995, and on-demand digital color printing was the focus of the years 1995 to 2000. During this period Web-enabled printing became imminent. This involved capturing Web images from the Internet and downloading them for printing on commercial printing presses.

The early twenty-first century has heralded the decline of prepress; the growth of variable data printing, e-commerce, and Internet publishing applications; and the practicality of virtual reality and "wetware" (the use of organic components of the body to grow microprocessors). By the year 2020 it is expected that nanotechnology will have taken hold with, for example, wires of one atom wide and memory cells composed of one electron.

What these technological transitions of over 500 years demonstrate is that the trend of significant technological change in shorter periods of time in the graphic arts is expected to continue. Digital on-demand short-run color printing is merely a stopping point on the way to the majority of printing being variable data and Web-enabled. These new technologies also have to go through the product development and marketing steps necessary for technology to evolve from concept to market demand and acceptance.

The following addresses some issues related to how graphic communication technologies change and offers strategies for effective implementation and marketing of the industry's technology.

Six Lessons from History about Technology

(1) Technologies tend to converge - One technology borrows from others. No single science has all the answers to solving the changing needs of society. For example, in graphic communication the technology of optics converged with the technology of emulsions to form the technology of photography. The technology of design has converged with the technology of digital computers to form integrated systems (one system that does all). Digital printing converged with the technology of the Internet to create Web-enabled printing.

(2) Change is typically evolutionary, not merely revolutionary - By the end of the 1990s, there has never been a technology in the graphic arts (with the exception of desktop publishing and on-demand digital printing presses) that has changed the posture of the industry within a five-year period. Prior to then, the traditional period from conception to implementation was 10 years. For example, it took approximately 20 years for the linotype machine to have an impact on the printing industry and approximately 10 years for phototypesetting to have an impact. However, since the 1990s the time it takes for a new technology to change the face of an industry has become greatly reduced. A few examples include cell phones, broadband Internet communication, digital photography, DVD media, flat panel computer displays, and cinema display high-definition television.

(3) Technology tends to expand existing markets and create new ones - Contrary to popular belief, technological innovation does not eliminate markets; instead it improves the ability to serve and expand existing markets and provides opportunities to develop new ones. For example, the development of the rotary offset press increased the speed with which printing could be produced and provided opportunities for printers to develop commercial markets in areas such as advertising. More recently, broadband Internet communication expanded desktop and laptop computer use by increasing the speed with which information can be sent and retrieved and made the home computer a common household and office device. Short run digital and variable data printing has not for the most part cannibalized printing jobs designed for traditional presses; instead it has provided opportunities to service both the short run digital and traditional printing needs of clients.

(4) Technology and practical needs share a mutual attraction - Technological change is rarely as disruptive as is often expected. For example, the advent of radio did not displace newspapers, nor did the advent of television displace radio or cinema. E-mail reduced but did not displace traditional mail and, likewise, cell phones reduced the need for but did not displace "landline" telephones. The World Wide Web did not displace commercial printing but provided new opportunities for print to support Web communication and visa versa.

(5) Results count and they are measured in practical human terms - Technology is often criticized for disrupting what is the norm at any given time. However, history shows us that technological progress cannot be stopped. It can be postponed or slowed down, but it eventually breaks through to satisfy human needs. For example, unions forestalled the introduction of phototypesetting for approximately six years between the mid 1950s and the early 1960s. But eventually the drive to satisfy human needs for efficiency and speed overpowered the desire to maintain the status quo. There were fairly large pockets of resistance to desktop computers, e-mail, and cell phones in their early years of general availability to the public. However, today even some of the most ardent opponents of such technology have come to realize that day-to-day functioning in the modern world is enhanced by their use.

(6) In the give and take between people and technology, people prevail - Technology must benefit people because people create and control technology to satisfy human needs. Technology does not control people, and people will not revert to old technology to fulfill practical needs once new technology is adopted. For example, many artists and designers resisted the thought that effective art such as painting and illustrating could be done on a computer. Yet once the new electronic medium was understood and perfected, many artists no longer resorted to using canvas or illustration board. Customers' expectations have also changed with new technologies—they want high-quality, low-price products delivered quickly, and this can no longer be provided using the "old" technology of the field.

To be successful in the marketplace, technological innovation must address more than just technology. It must address the human interface to all technology. Questions that help determine if a technological transition is ready to take place include:

- Does present technology no longer serve a need?
- Are people ready for a new technology to the point where they will invest in it?
- Will a new technology be embraced as a means to make work easier, but not less important, and as a way to better service market demands?

The most successful new technologies are those that present a proven paradigm shift in the form of a device or concept that helps solve old problems in a new and better way, or the technology aids in understanding ideas, systems, and behaviors. Some examples of concepts that have led to paradigm shifts include Charles Darwin's theory of the "survival of the fittest," which taught us that, in nature, only the strong survive; Sigmund Freud's theories of psychoanalysis explained human behavior as being a reaction to early childhood experiences; Ivan Pavlov's study of conditioned reflexes explained human motivation; John Watson's explanation of behaviorism helped us to understand the influence of one's environment on behavior; and Thomas Kuhn's "Structure of Scientific Revolutions" taught how old research methods are replaced by new and improved ones. As for paradigm shifts that focus specifically on technology, the combined work of Dorothy Leonard-Barton of Harvard Business School and William Kraus of the General Electric Corporation has resulted in a new paradigm for "Implementing New Technology" which teaches that there is a natural resistance to new technology and technological change and that this resistance must be addressed.

Implementing New Technology

According to Leonard-Barton and Kraus, there is a gap between the value of technology and the ability to put it to work effectively. This means that in many cases a large investment may have been made in a new technology but the technology is subsequently not used to its maximum potential. Examples in the graphic communication industry include the acquisition of high-end integrated systems and very expensive and powerful computers. A company may acquire a high-end system and use it nearly exclusively for one or two applications. This occurs for a number of reasons. One is that the company was attracted to the glamour, or what is sometimes called the "magic," of the system and has every intention of using its capabilities to expand markets and attract new ones. Yet this rarely happens as the company continues to handle the day-to-day routines of serving the traditional needs of its clients and meeting deadlines. Sometimes the problem is resistance from those who are expected to operate the new technology; they may not be properly prepared for it.

Yet another problem that graphic communication companies face is overcapitalization in order to bring in new technology. This approach makes sense if the company expands markets and generates additional revenues from the new technology. However, this does not always happen, and companies that have made large investments to attract and keep customers soon find themselves out of business.

A study conducted in northern California shortly after integrated electronic prepress systems were introduced to the printing industry confirmed that print buyers are attracted to printing companies that appear to be modern. The study was conducted to address the concerns of northern California—and particularly San Francisco Bay area—printers that much of the printing generated in that area was being produced by printers in Los Angeles or Portland, Oregon. The study found that Bay Area printers were stifled by strong union constraints on bringing in new technology, whereas union influences were not as strong in the other two cities. Los Angeles and Portland printers embraced new electronic prepress technology soon after its availability, while San Francisco printers retained traditional operations. The results of the study further indicated that, with all else being equal—e.g., price, quality, and speed of delivery—given the choice of purchasing printing from a printer having a modern-looking plant or one having old equipment, most printing buyers would purchase from the modern-looking plant.

There is a distance between technological promise and achievement. This raises the question: Does technology do what its advertising claims it does? Advertising tends to exaggerate the ease with which technology can be maximized. Advertisements do not address the cost of operating technology, the life of components, the need for maintenance contracts, the cost of supplies—all of which must be taken into consideration when determining whether a given technology will be cost effective. For example, nearly all major electronic printing equipment requires that the owner of the equipment purchase a service contract; these contracts are usually expensive. However, the user has no choice because only the company offering the service contract has the specialized skills needed for such repairs. This is rarely noted in product advertisements or brochures—nor is the cost of supplies or replacement parts. In today's market, the investment in hardware often pales in comparison to the ongoing costs of service and consumables such as inks and toners.

Leonard-Barton and Kraus list six requirements for implementing new technology successfully—a dual role, a marketing perspective, a framework for information, multiple interval markets, an understanding of promotion versus hype, and a "risky" site that provides "safe" innovation.

Dual Role - Because the user of new technology is not always ready to use it, the developer or manufacturer of the technology must implement it at the plant where it is being installed. With very sophisticated technology such has high-end integrated systems or large electronic printing presses, the manufacturer will often put a technician in the plant for an extended period of time; it could be months before plant personnel are ready to operate the technology unassisted. Examples abound where such technicians were "pirated" away from developers at high salaries.

Marketing Perspective - An intelligent marketing approach to ensure sales of new technology is involving users in the technology's design phase. This boosts user satisfaction because the user has become familiarized with the technology prior to purchase. An innovation's technical superiority alone will not guarantee acceptance in the marketplace. The best technology in the world will not reach the market if prospective buyers do not understand and feel comfortable with it. Hence, the wise system developer will include some of its best prospective customers in planning sessions.

Additionally, prior to the installation of new technology, it must be marketed to operators. Ideally, operators should be convinced that the technology is most suited to meeting the goals of the company, and operators should be included in the decision to acquire it. The management of one major printing company in Canada sends operators to non-competitive plants in Canada and the United States to see a given technology at work and how peers are using it. Another company sends operators to expositions where new technology is being exhibited and to manufacturing plants where it is being built with the hope that the operators will recommend purchase. One theory is that if operators endorse it, the likelihood that the technology will operate successfully and profitably is greatly increased.

Framework for Information - Ideally, one person should coordinate the work of gathering all of the information needed for the implementation of new technology. That person should observe current job routines and workflows and identify bottlenecks that need to be eliminated. An analysis also should be conducted to ensure that the new technology will not create bottlenecks, as in the case of the production of one department increasing to the point where the next department will be unable to handle the increased volume.

This latter problem occurred in the graphic communication industry with the automation of prepress in the early 1980s. Graphic arts photography via scanning, image assembly, and platemaking started to occur so rapidly that press departments could not handle the products of prepress fast enough and bottlenecks occurred. Hence, the next focus of automation was in the press department, and the late 1980s and early 1990s saw tremendous improvements in press speeds through the introduction of electronic press controls. By the early 1990s, sheet-fed and web press speed nearly doubled and wiped-out the bottleneck between prepress and press. However, the elimination of this bottleneck caused another one in binding and finishing departments, where printed sheets or printed web rolls had to be slit, scored, folded, collated, trimmed, and so on. Thus in the mid-1990s great improvements in binding and finishing speeds were developed, particularly in the form of online and integrated press and finishing functions, and this reduced bottlenecks in the bindery.

Attention also must be paid to the sections of work that require user decisions about tools and materials; safety and health issues apply here. New technology sometimes raises legal issues related to employee safety and environmental protection. Someone must take responsibility to ensure that all Occupational Safety and Health Act (OSHA) and Environmental Protection Agency (EPA) regulations are being met with the implementation of new technology. In the printing industry, exposure to chemistry and volatile gases and waste disposal are grave concerns. Some technology requires chemicals that must be appropriately shielded from machine operators, and such protections are typically very costly. Disposal of unused inks, fountain solutions, photographic chemicals, plates, and related waste also must be monitored. Without appropriate protections in these areas, successful implementation and operation of new technologies may not be possible.

A common occurrence in the graphic communication industry is a case where technology works at one company with no problems whatsoever but a second company using the same technology has ongoing difficulty with implementation and operation. In most cases the equipment and supplies provided by the major developers and manufacturers are reliable. However, for seemingly indefinable reasons, the employees of one company, for instance, will swear by the superiority of a particular printing plate, but another company doing the same type of work on a similar press cannot run that brand of plate without encountering problems. The same situation happens with supplies such as inks, fountain solutions, and so on; and to equipment such as scanners, exposure units, computer-to-plate devices, and presses. Hence, if the technology itself is reliable, there must be something else causing inconsistency of performance from one company to the next. Employee attitudes about the technology, how they are introduced to it, their personal relationships with equipment and supply vendors, and what they hear from others all play a role in the motivation to implement and operate technology successfully.

The person responsible for coordinating the work of implementing technology should also investigate how manufacturing processes within the company relate to each other and determine the extent to which machine operators are dependent on materials, personnel, maintenance, and so on. New technology sometimes changes technical relationships and dependencies between departments or cost centers. Such changes could provide improvements or hindrances, and companies replacing old technology with new should sufficiently study the impact on workflow to avoid surprises. For example, supplies have to be at the right place at the right time to continue a smooth flow of production. A question to ask before implementing technology is: Will changes in technology impact supply availability?

The graphic arts industry is particularly reliant on a systems concept that ensures flow of production in a way that minimizes downtime, waste, time, and cost. Industry segments such as commercial printing are extremely competitive and rely on low profit margins to obtain and retain customers. The smooth and expeditious running of most printing plants relies on a rigidly coordinated effort between all cost centers of production. For example, the efficiency and quality of work produced in prepress influences the ease with which work can be performed properly in press departments. Likewise, the quality of work performed in press departments influences how efficiently work can be produced in the binding and finishing department.

More specifically, in the traditional flow of printing production there are six cost centers, as noted earlier: art and copy preparation; graphic arts photography (though nearly obsolete) including scanning; image assembly; platemaking; presswork; and binding and finishing. While there may be fewer cost centers in highly automated companies, the concept still applies. Typically, the earlier stages of production represent lower cost centers. In other words, it cost less to perform tasks in art and copy preparation departments than it does in graphic arts photography, image assembly, or platemaking. This has nothing to do with the abilities or qualifications of the people who work in these areas, but relates to cost of labor (number of people working in these areas) and the cost of technology required in each cost center. Art and copy preparation departments, even if heavily automated with computer systems as most are today, require fewer people and less expensive equipment than other prepress departments of printing plants. Likewise, press departments and binderies are relatively labor and capital intensive, and require more expensive equipment and supplies. Printing presses often cost millions of dollars, and paper, the most costly disposable commodity in printing, typically represents 35 percent to 50 percent of the cost of printing. In other words, if it cost a client $2 million to purchase a printing job, it is not unusual for approximately $1 million of the cost to go toward paper. With this reverse pyramid of cost, the "system concept" applied to printing production teaches that work should be organized in such a way that as much work as possible is performed in the lower cost centers, thus minimizing the time it takes to produce work in the higher cost centers.

One goal in this system is to avoid backward movement, i.e., having to return work to earlier cost centers. This could happen, for instance, with poor imposition, or improperly prepared plates that are not discovered until a job is already on press. Another example is when files are not prepared properly in a company's IT department. Such breaks in the flow of production often mean the difference between profit and loss in the printing industry. One of the costliest interferences in production flow is when paper becomes unstable and stretches or shrinks due to relative humidity (RH) changes around the press and in the paper; this will adversely effect color registration and paper movement through sheet-fed and web presses. Changes in moisture content can occur not only from improper atmospheric conditions, but also from heat changes caused by printing presses and by moisture added to paper during press runs. An important question to answer prior to installing a new printing press is: What changes will be necessary with regard to paper handling conditions?

Multiple Internal Markets - The higher the organizational level at which managers define a problem or a need, the greater the probability of successful implementation of new technology. When top management shows genuine concern that a new technology must solve certain defined problems and that the technology must be made available within a defined but reasonable cost range, developers will focus on meeting these production and bottom-line needs. All too often, for instance, printing equipment manufacturers develop new technology without addressing the focused needs of individual potential users. The problems that the new technology addresses—such as faster speeds, higher quality, smaller "footprints," safer conditions, and so on—are broadly defined by the developer and marketed to companies that are experiencing those problems. With this comes a price tag determined solely by the developer with the hope that the price is acceptable to the industry in general.

On the other hand, a company seeking new technology to solve problems specific only to itself is best served when top management approaches developers with the request to develop the technology or modify existing technology. When top management takes on such an exploratory role, this is evidence of the seriousness of the need for and the likelihood of contracting for the technology prior to creating it.

Promotion vs. Hype - Over-selling a system can be dangerous to implementation and successful use of new technology. There is a gap between the perception of what a technology promises to do and the reality of what it can actually do that must be closed. People develop negative attitudes toward technologies that do not work for them; this applies to equipment and to supplies. Leon Festinger, a well-known communication theorist and researcher, explained the relationship between attitude and perceived credibility through his famous Theory of Cognitive Dissonance. The theory teaches that "knowing inconsistency" breeds negative attitudes that take a long time to reverse, and that "knowing consistency" breeds positive attitudes that must be continually reinforced. For example, if a printer purchases a computer-to-plate system thinking that it is the best technology for solving a multitude of production problems, and after the purchase and installation learns that the system's abilities were exaggerated, the credibility of the manufacturer plummets. It then may take years of consistent positive image building and product performance for the manufacturer's credibility to be restored. Hence, when one lies, misleads, or hypes, credibility drops. It then takes a long period of consistency in what one says or promises, and what one does or delivers, for credibility to reach its previous high. Honesty represents an under girding for doing business in the twenty-first century.

Risky Site—Safe Innovation - A pilot site helps to iron out potential implementation problems and raises questions that must be answered prior to marketing. Testing new technology at a sub-normal performance site answers most questions about implementation. This is a particularly interesting concept and anomalous to what one would expect. Very often, new technology is tested in a beta site representing pristine and sometimes antiseptic conditions, e.g., a clean environment, excellent lighting conditions, and well-trained operators. However, this tells nothing about how the new technology will work under typical operating conditions. Hence, the recommendation is that, for example, beta testing a new printing press in a plant having relatively poor environmental conditions and minimally trained operators will tell more about the functioning of the new technology than would testing in an ideal environment. If the problems discovered under poor conditions can be resolved, then the technology is likely to work well in any environment.

As a further illustration of this concept, a major manufacturer of a digital "printing press" donated one to a university to use in educating its students, but was appalled to learn that the plan was to install it in an immaculate, antiseptic, computer laboratory. The market-savvy management of this company wanted the system installed in a traditional printing press laboratory next to conventional presses. They felt that students should learn, and industry visitors to the university should see, that even though this new technology looked quite different from the conventional presses, it could compete with them and could operate effectively in a traditional press environment.

The principles and concepts described in this chapter relate to all facets of the graphic communication industry, including service providers and equipment and supply manufacturers. The next chapter explores the various segments of the industry.

5

Graphic Communication Industry Segments

The graphic communication industry is comprised of 19 segments:

1. Commercial Printing
2. Newspaper Printing and Publishing
3. Magazine/Periodical Printing and Publishing
4. Book Printing and Publishing
5. Business Forms and Bank Stationery Printing
6. Financial and Legal Printing
7. Greeting Card Printing
8. Yearbook Printing
9. Folding Carton Printing
10. Flexible Packaging
11. Corrugated Box Printing
12. Metal Decorating
13. Label Printing
14. In-Plant Printing
15. On-Demand (Quick) Printing
16. Prepress Vendors
17. Service Bureaus
18. Digital Variable Data Printing
19. Print Broker

Commercial Printing—Commercial printing is comprised of approximately 45,000 establishments in the United States, thereby making it the largest industry segment with regard to number of establishments. This segment performs general printing and commercial printing companies must be equipped to produce a wide variety of products from simple letterhead to complex four-color printing. Most jobs are relatively short run, and approximately 80 percent of all commercial printers employ fewer than 20 employees.

The commercial printing segment is highly competitive and relies on high volume and low profit markup to sustain itself. Being able to provide a diversity of services in prepress, press, and post press and having access to a broad range of supplies such as paper and ink are crucial for commercial printers to remain competitive. With a focus on service, this segment services local needs geographically, though the larger commercial printers serve regional and national accounts. As is typical of most industry segments, the rapidly growing area of commercial printing is in color reproduction and digital printing.

Newspaper Printing and Publishing—The segment is defined as that which prints on newsprint including all types of newspapers, advertising inserts, and special publications. While daily newspapers are on the decline in the United States, and have been for many years, weekly, community, and special interest newspapers are growing. The trend in newspaper printing and publishing is toward greater use of color, which was spearheaded by the Gannett Corporation's *USA Today* in the early 1980s. Since then nearly all major metropolitan and many community newspapers have introduced color editions. Some recent technological advances are in the development of "non-rub" inks. Such water-based inks have been successful in newspapers produced with the use of the flexographic printing process, and research and development shows that such inks will become suitable for other processes as well.

Some newspapers have experienced revitalization and growth resulting from media mergers and acquisitions. This is a trend resulting in media conglomerates owning two or more media such as newspapers, radio, broadcast and cable television, and movies. Commercial access to satellite transmission has resulted in a number of newspapers providing national and even international editions. However, such growth will continue slowly because local news and advertising are still the main draws for readers.

Moving rapidly into Internet and World Wide Web applications, local newspapers are beginning to diversify their services into electronic databases. These databases supplement the printed versions of their products in editorial content and classified advertising, and in telephone and facsimile services providing automobile, employment, real estate, and other similar listings.

Magazine/Periodical Printing and Publishing—The trend in magazine and periodical printing and publishing is toward specialty publications and has been for many years. The general interest magazine has given way to magazines and periodicals focusing on special interests. There is a magazine for nearly every interest: news, sports, health and fitness, travel, fashion, home decor, and so on. Occupations also have their own periodicals, and the wide acceptance of desktop publishing has resulted in more titles with smaller circulation than ever before.

The major trend in this segment is toward more use of color and highly sophisticated integrated front-end (prepress) systems by some of the larger publishers (*Time, Newsweek, U. S. News and World Report*) and some national consumer magazines. While there are more publications with smaller circulations, the collective circulation of all magazines and periodicals continues to increase in spite of postal rate increases. This shows that magazines and periodicals still are desired by a public with other media to choose from, including the Internet. The industry segment is one that focuses on high-quality coated paper coupled with good-looking color. Continued improvements in color reproduction and distribution methods, and the ability to produce regional or "demographic" editions, keep magazines and periodicals competitive with television and the Internet.

Magazine publishers are also taking advantage of technology that allows demographic printing and distribution. Using this technology, magazines provide advertising and even editorial content of interest to readers of a particular geographic region. Some publishers offer split runs and deliver by zip codes, and selective binding is being used for personalized advertising. Magazine production today is electronic in the preparation of layouts, pagination, and image assembly. This industry segment will continue to grow as a result of the increased availability of low cost and sophisticated desktop publishing systems.

Book Printing and Publishing—Book printing and publishing has been a slow growth area for a number of years. Strained school budgets are keeping textbook purchases flat, despite growth in elementary and secondary school enrollments. Aside from textbooks, it appears that people do not read books for enjoyment as much as they did in the past. Competition from other media such as television, DVDs, the Internet, and video games have influenced the degree to which people read books.

A growth area of the book printing and publishing segment is the production of manuals and technical books, which are necessary to keep people apprised of operational procedures for rapidly changing technologies. This industry segment services primarily a black-and-white market using uncoated paper printed on standardized sheet-fed perfecting presses. Any modest growth in book publishing generally should be accommodated by moderate cost increases as paper mills and book printers become more efficient in the manufacture and printability of recycled, acid-free, and alternative paper sources.

As is the case with newspapers, the book printing and publishing industry is frequently part of consolidations and mergers across media. Growing competition from computerized information systems such as CD-ROMs and DVDs is forcing book printers to diversify, and new technologies are influencing how they create their products. The interface of sophisticated typesetting/pagination software to LED (light-emitting diodes) printers provides opportunities for customized textbooks for college classrooms. This process also enables publishers to provide limited-edition textbooks geared toward regional preferences. Related to this trend are simplified front-end systems coupled with high-speed presses that provide opportunities for high volume "overnight" production and dissemination. Desktop publishing and on-demand printing enable printers to easily make corrections and fill orders quickly. The availability of low-cost typesetting and design technology is easing the entry of new publishing firms into an increasingly competitive market, and this is expected to push new book title production to greater levels in the twenty-first century.

The publishing of textbooks and general hardcover and paperback books takes place worldwide. Globalization of this industry segment will grow as the technology for transmitting and receiving editorial and related copy from any place in the world becomes common to printers and publishers. Technology has also made it cost efficient to produce short runs of books, which has reduced the need for high-volume inventory and storage of books.

Business Forms and Bank Stationery Printing—With the proliferation of electronic funds transfer systems (EFTS) and automated teller machines (ATM), bank stationery printing as a separate industry segment is being eliminated, and bank stationery products still in use are typically produced by business forms printers. Business forms printers usually produce three main products: snap-out or unit set forms, computer or continuous forms, and specialty forms. Snap-outs, which have become popular for utility invoicing such as water, electric, and natural gas bills, include a mailing envelop, an invoice, and a return envelop all gathered in one unit. Computer forms are typically those used for inventory control, classroom rosters, purchase histories, and so on. Specialty forms such as sales books comprise a very small and declining portion of business forms products.

Business forms printers may also produce checks, bankbooks, passbooks, deposit and withdrawal forms, and related items. Checks, with the exception of Traveler's checks, represent the largest portion of today's bank stationery products.

For many years the business forms segment has been highly profitable, mostly because it manufactures specialized products on standardized equipment with a relatively small variety of supplies. Its small format and narrow web presses are relatively simple and require low operator skill levels; therefore, operator wages are not as high as in other industry segments. Business forms products are typically produced on uncoated paper and, while the use of color is growing, there is still little color used on business forms.

While the use of printed business forms is on the decline, an interesting growth area is in printing personalized forms with information directed to the individual recipient. For example, some utility bills such as gas, electric, telephone, and cable now carry personalized messages and advertising directed to the addressee. Large companies using variable data printing equipment and sophisticated databases usually produce such business forms.

The business forms industry segment typically serves local regions with some companies serving a national, but not international, market. Business forms printers have moved rapidly into electronic and desktop publishing as forms printing has moved online. Therefore, traditional business forms printing is on the decline and may cease to exist as society adopts electronic alternatives to printed forms.

Financial and Legal Printing—This segment produces materials for borrowing money and for money transfers, as well as materials for informing the public and stockholders about transactions. Products include stock certificates, prospectuses, lottery tickets, registration statements, loan coupon books, legal briefs and documents, Traveler's checks, bonds, leases, Security and Exchange Commission (SEC) filings, proxy materials, foreign currency, and related certificates and documents. Financial and legal printing is typically national, but there are some companies that serve local and regional needs. Currency is printed by governments or by contract to financial printers. Until recently, much of the world's currency was printed in the United States for developing nations concerned with security and quality control. However, with advanced technology becoming increasingly available, even to developing nations, financial printing is rapidly becoming a global industry.

Since the stock market crash of 1987, this industry segment has experienced consolidations and liquidation, and automation in the form of electronic data transfers has streamlined the process of producing and distributing financial and legal documents.

Financial and legal printing is highly specialized with only a few companies controlling approximately 90 percent of the volume. One reason for this is that the SEC controls financial and legal printers; this is the only printing industry segment that is largely controlled by the federal government. Such control is necessary to protect the security of the information contained in many financial documents, particularly that which relates to the buying and selling of companies and stock trading. Because of the importance of security and confidentiality, having loyal and honest employees is important. Indeed, employees of this industry segment must often have SEC clearance.

Products in this segment typically are printed on uncoated paper and there is little use of color, although image detail is often intricate for security purposes. What is of utmost importance in financial and legal printing is the accuracy of information contained on documents. Security of financial documents necessitates complex technology and creative anti-counterfeiting techniques such as heat-sensitive inks, holograms, watermarks, special papers, and security threads running through currency. The financial and legal printing industry segment is highly specialized and often highly profitable. Financial printing will continue to be dominated by a few companies capable of maintaining security of highly confidential information and accommodating very short delivery times through databases and direct digital printing.

Greeting Card Printing—The greeting card printing segment is highly specialized, innovative, and creative. It could be highly profitable and has a unique focus on quality. This growing printing industry segment is controlled by a very few companies. In fact, three companies produce approximately 80 percent of all greeting cards in the United States, with Hallmark having the dominant share of the market. Hallmark's focus on quality has become the benchmark for quality not only in the greeting card industry but in other printing industry segments as well. Hallmark imposes rigid standards on its printers and has one of the most demanding quality control and quality monitoring programs in all of the graphic arts. The company requires that in mass production of the same card, each card must look exactly alike with no color variations. The consumer will notice no color shift from one card to the next on a retail store rack.

The products of the greeting card industry segment are produced on coated and uncoated paper and on specialty substrates including plastics and foils. Unlike general commercial full-color printing, the color requirements of greeting cards often demand the use of six-, eight-, and sometimes ten-color presses to print not only ink but special metallic pigments, laminations, and coatings. The products of the industry also often require specialty processes such as embossing, die-cutting, and holography applications. A current trend is to manufacture regional or "demographic" cards that are identified with a particular community, city, or state, and personalized cards on which the consumer can interject a personal message or even a picture. It is of interest to note that the advent of electronic greeting cards on the World Wide Web has had little impact on consumer demand for printed greeting cards.

Yearbook Printing—Yearbook printing is a segment that focuses on the specialized products purchased annually by nearly every high school and college in the United States. With their standard format, yearbooks lend themselves to highly routine production procedures in the way copy is prepared and handled at the printing plant and the finished product is distributed to the schools. It is an industry segment with a relatively unsophisticated client base—high school students and young adults. The emotional appeal of the product to its buyers far outweighs concerns for quality and maintaining tight production and delivery schedules.

Nearly all copy preparation is done by students at the schools in accordance with instructions for layout, type, and picture-use that are provided by the yearbook printer. Because of this, copy and particularly pictures often lack technical quality. In the printing plant, production tolerances are relatively wide and inexperienced practitioners are often hired and paid wages that are lower than their more skilled counterparts who work in industry segments where greater efficiency and quality are mandated by customers and competition. In yearbook printing, production workers are sometimes hired away from other low paying jobs on a part-time or seasonal basis when yearbook production is at its peak. The main delivery requirement of most customers is that the product be delivered anytime before spring graduation, usually in May or June

Products in this industry segment are generally printed in black and white, but market forces have increased the degree to which full-color reproduction is required. In response, some yearbook printers have upgraded their technology and improved their products. There are relatively few companies that specialize in yearbook printing. With its standard equipment and procedures, low investment in quality assurance programs, and relatively low wage base, the yearbook printing industry tends to be highly profitable.

Packaging—Package printing is a large part of the graphic arts industry, with gross annual sales revenue nearly equal to all general commercial printing in the United States combined. There are five packaging industry segments: folding carton printing, flexible packaging, corrugated box printing, metal decorating, and label printing. The component that runs through each of these segments (although to a lesser extent in corrugated box printing) is the importance of quality consistency. Packaging companies, particularly those that produce packages with pictures of food products, flesh tones, and other natural colors, realize that the human eye is most sensitive to natural colors. Thus, people are reluctant to purchase products whose packaging contains an unappetizing photo of food. In the case of products that are placed on the skin such as cosmetics, photographs of people must be appealing and natural. Therefore, quality assurance and quality control are vitally important in nearly all phases of package printing—sometimes more so than in the greeting card industry.

Environmental concerns are uppermost with package printers, particularly those with heavy color printing. They are under pressure to develop methods of effective printing on easily recyclable materials.

Some companies specialize in package printing, while others do package printing on a contract basis with product producers. Much package printing is done by product manufacturers, who often find it more efficient and economical to develop their own package printing facilities than to contract out such work.

Package printing is expected to experience heavy growth in the years ahead because it is the only printing industry segment that is not negatively impacted by the Internet and World Wide Web. In fact, many digital printing press manufacturers are developing presses to produce short-run packages and packages with variable data information on them.

Let's take a closer look at the five packaging industry segments.

Folding carton printing occurs on heavy substrates and boards that are easily folded into a permanent form. Cereal boxes, detergent boxes, and many cosmetic boxes represent this category of package printing. Package engineers are usually involved in the design of folding cartons to ensure stability, durability, appropriate display characteristics, and storability of cartons. The substrates used sometimes have a liner and are typically highly coated. Folding carton printing is normally a sheet-fed process using standardized but relatively large four-, five-, and six-color printing presses. The fifth and sixth units of these presses are used to add additional colors beyond the four process colors and to add varnish, or for other finishing applications to enhance the visual appeal of the printed carton. Folding carton printers are often owned by large paper companies or by the producer of the product. Paper mills sometimes see diversifying into the folding carton printing business as a lucrative opportunity to convert their main product, paper and board, into usable and marketable products. Food manufacturers such as Kellogg's and General Foods enjoy certain efficiencies and enhance their control over product manufacture and distribution by also controlling package manufacturing. In essence, this industry segment is made up of a small number of large printing and converting companies that do the majority of folding carton printing in North America.

Flexible packaging usually involves the use of substrates other than paper and board—mainly foils, plastics, cellophane, and other pliable materials on which a printed image can be produced. Such materials have little or no ink absorption and extremely high ink holdout, and are typically printed using processes other than lithography. The main processes used in flexible package printing are gravure, flexography, and, to a much more limited extent, screen printing. The high solvent content of conventional gravure and flexographic inks enhances the bonding quality of the inks to the substrate and the overall image appearance. Conventional lithographic inks, designed to carry a small amount of water and to absorb into porous substrates, do not lend themselves to non-absorptive substrates such as plastics and foils because of diminished color brilliance and the tendency to easily scratch or rub off. Flexible packages are usually produced in extremely high volumes for major product manufacturers and are printed on high-speed web presses typically used in gravure and flexographic printing. Unlike lithography, the matter of ink drying is more complex in gravure and flexographic printing, particularly when printing on high ink holdout substrates. Drying must occur after each color is printed and must be instantaneous to avoid smears and loss of image gloss. Therefore, highly complex and expensive drying methods are used such as infrared, ultraviolet, and electron beam dryers. These dryers are housed between each unit of the press to dry each color immediately after application to the substrate.

Corrugated boxes are typically made of low-quality and highly absorptive craft paper and board with irregular and non-white surfaces that do not lend themselves to high-quality printing. The typical packaging carton with its thick sides (top and bottom and liner) is intentionally made highly compressible for protection of the items stored in the box. The thickness of the substrate and its compressibility are attributes that make corrugated boxes poor printing surfaces, and any printing found on such boxes are usually in one color—black. Flexography is the preferred process for printing on corrugated boxes. In cases where product manufacturers and distributors desire attractive and well-printed graphics, they usually prepare and print such graphics on large wrappers or labels which then are applied to the outside of the corrugated box.

Metal decorating is printing on aluminum, steel, and other hard metal surfaces. While foil printing can be considered a form of metal decorating, it is usually not included in this industry segment and instead is classified as flexible packaging. Metal decorating is perhaps the most specialized of the packaging segments. Metal can be printed in sheets on sheet-fed presses, in rolls on web presses, or on specialty presses that print the metal after it has been formed into a container. This latter process is the way soda and beer cans are printed. Lithography is the main metal decorating process, although flexography and screen printing are sometimes used. Screen printing is popular for printing pre-formed containers. Metal decorating requires highly specialized inks that must dry instantly, adhere to metal, and resist scratching.

Label printing refers to printing that will be applied to an already formed container made of any substrate or material including board, foil, metal, plastic, and glass. All major printing processes can be used to produce labels, although lithography and flexography seem to be the two most popular processes in this industry segment. When lithography is used, printing takes place on large sheet-fed presses where many labels can be assembled or stepped and repeated for printing on a single sheet. The backs of the sheets receive any necessary adhesive and are then die cut. Flexographic label printing usually occurs on narrow-width flexographic web presses, many of which do on-line finishing such as die cutting scoring, removal of paper surrounding the die cut labels, and slitting. The substrates used in this process are often pressure sensitive with a peel-away backing applied prior to printing. There are numerous label printers in the United States, some of whom use four-, five-, six-, and up to eight-color presses. An understanding of adhesion and gluing requirements is important; the most beautiful label has no value if it does not adhere. Different substrates require different adhesion properties. Weather conditions and the storage environment influence the effectiveness and longevity of labels. Typical label products include package labels, pre-glued labels, dry gum labels, pressure sensitive labels, self-adhesive labels, flat labels, roll labels, bar code labels, flat wrappers, heat-sealed wrappers, and glued wrappers.

Other overall considerations that relate to most of the packaging printing industry are maintaining consistent quality, working with standardized formats for sheet-fed and web presses, understanding and working with many and varied substrates, and using highly specialized inks that are rub resistant and dry quickly. Two other broad considerations relate to Federal Drug Administration (FDA) requirements and the use of the Universal Product Code (UPC). The FDA has placed some regulations on package printers, particularly where food packaging is involved. There are regulations regarding ink formulations that have solvents and volatile and gaseous emissions that can contaminate food. There are also regulations regarding the structure and material of package liners that separate the food product from the container. Packagers must prevent contamination from emissions that may penetrate through the container substrate, as well as retain the freshness of the food within the package.

The UPC was devised in the early 1970s with the immediate goal of speeding the supermarket checkout process and the long-term goal of realizing the "shelfless" supermarket. The UPC—made of numbers, letters, and vertical lines of varying thickness and spacing—can represent millions of bits of information. Programmed into the computers that read the UPC is information such as price, inventory, location in store, discounts and sales, age of product, how long the product has been in the store, how rapidly the product is selling, and so on. Today, the symbol is used primarily for price control and checkout scanning.

The key to the successful application of the UPC is its printed resolution. Standards have been established for acceptable resolution and if printers do not maintain these standards, the symbol will not read properly when scanned in the supermarket. The need to include this code on packages during regular printing has resulted in an overall improvement in package printing. As presses had to be recalibrated for the effective printing of the UPC, printers found that they significantly improved the resolution and quality consistency of the package itself.

One of the latest technologies that may impact the packaging industry is RFID (Radio Frequency Identification). With RFID, each package has a built-in chip that allows electronic and automatic alteration of package information such as contents, ingredients, price, and so on. RFID technology also allows many products in a cart to be priced with just one scan of the cart.

In-Plant Printing—The in-plant printing segment consists entirely of "captive" plants servicing one client; one plant services the specific needs of a parent company. In-plant printing establishments range from thousands of employees to only one, but most are small and employ several people. Very few employ over 100 people. Examples where in-plant printing operations are found include manufacturing companies that have an on-going need for brochures and manuals, company reports, and newsletters; universities that have a continuous need for department publications, reports, and course materials; insurance companies and financial institutions that must regularly publish reports for customers, policy holders, and investors; and government agencies.

Products range from simple business cards and letterheads, business forms, sales brochures, newsletters, technical reports, and direct mail advertising pieces to fancy and complex four-color annual reports. The largest in-plant printing facility in the United States is the United States Government Printing Office (GPO). Employing more than 3000 people, the GPO is responsible for producing the daily Congressional Record. The GPO also prints thousands of other items ranging from simple reports to fully bound hardcover books published by the federal government and its various agencies. Kellogg's and General Foods have their own in-plant facilities to produce the packages in which their products are placed. Westinghouse Corporation has a large in-plant printing plant to produce most of the manuals, reports, and some of the advertising required by Westinghouse.

Larger in-plant facilities could be equipped with some of the most sophisticated printing equipment available anywhere. Smaller facilities, representing the vast majority of in-plant establishments, are typically equipped with relatively simple and small format equipment to produce uncomplicated, routine items needed by the parent company.

Several decades ago, in-plant printing experienced significant growth through the development of small offset duplicators and presses and simplified platemaking processes. Recently there has been rapid growth of this industry segment with the increased availability of desktop publishing opportunities, electrostatic and inkjet printers, and "turn-key" electronic digital operations. With the "miniaturization" and cost reduction of such equipment, companies that previously relied on the services of a commercial printer now find it economical to acquire digital equipment and produce their own printing. In-plant printing facilities are generally popular in government, education, insurance, food processing, large manufacturing, and health care organizations.

On-Demand (Quick) Printing—On-demand printing is often called quick printing, and it represents one of the newer industry segments. Establishments in this business offer walk-in storefront service and provide printed products immediately through sophisticated desktop computers and electrostatic printers. Such establishments specialize in small volume and simple formats, and also typically provide simple binding and finishing services. The products of on-demand printers most often are produced on uncoated paper and printed with black ink. However, on-demand color printing has grown rapidly in recent years. Some companies provide color services through color copiers.

One trend in on-demand printing services allows the customer to participate in production. This is done through desktop publishing computers that the customer uses for a fee. In this manner the customer becomes the producer and controls all prepress operations. From there, the copy is then produced on a high-speed electrostatic copier. These copiers have replaced small format offset duplicators as the printing process in this convenience segment of the printing industry.

In 1969 there were approximately 1,100 quick printers, also known then as "copy shops," in the United States. Today there are nearly 40,000 establishments in the on-demand printing business. The majority of them are individually owned single establishments, and about one-fifth of the companies have more than one location. A growing number of on-demand printing establishments are part of franchises. Franchise printers benefit from discounts provided by nationwide group purchasing programs and periodic training programs in marketing, sales, and production. While nearly all on-demand printing serves local communities, some larger franchise printers operate internationally through sophisticated computers for the transmission and receipt of copy.

Prepress Vendors—Prepress vendors have traditionally been referred to as "trade shops" that provided prepress services to printers that were not equipped to do prepress work. Services included composition and imaging; line, halftone, and color separation negatives and positives; electronic prepress services; photoengraving; and platemaking. This industry segment was revived in the 1980s due to the availability of sophisticated integrated color electronic prepress systems (CEPS) that provided efficiencies in productivity and quality that conventional prepress technology was unable to provide. Basically, one CEPS replaced the need for separate departments for typesetting, graphic arts photography, image assembly or stripping, and in some cases platemaking. Unfortunately, these systems were expensive. Therefore, to enjoy the benefits that CEPS offered, printers returned to purchasing services from prepress vendors who had invested in integrated systems. The vendor justified the investment by providing prepress services simultaneously to numerous printers locally or nationally.

Through improvements in CEPS, the availability of low-cost color desktop publishing systems, and the increase in the number of prepress vendors providing such electronic services, color had become less expensive and printing color "on-demand" had become common. Prepress production was simpler, and there was less labor needed. Standards have been developed so that electronic prepress systems from different manufacturers can be used together. This industry segment has experienced acquisition and merger by larger prepress vendors, which in turn gave them access to expensive CEPS equipment. Today prepress vendors are on the decline as prepress technology has been simplified and miniaturized by electronic digital equipment.

Service Bureaus—Service bureaus is the one industry segment that fully acknowledges the printing industry's place in electronic printing and publishing and is committed to using desktop publishing and other electronic applications in preparing and reproducing copy for printing. A service bureau typically works from computer files, a modem, or other high-speed delivery devices. It is not unusual for such establishments to have no typesetters or cameras on the premises. Most service bureaus will only accept "electronic copy" and will not accept traditional paste-ups or mechanicals. Some will output computer files to resin coated (RC) paper, film, or four-color separations. Many provide electronic scanning services, slide imaging services, and inkjet or other digital proof output. They also provide black-and-white line art, halftones, and color separations; electronic trapping and masking; high-resolution scans of color photography and illustrations; and large format digital prints. With today's direct imaging printing presses, service bureaus can send electronic files on disk or by modem to a direct imaging press where plates are electronically imaged within 15 minutes. The press can be at a remote site and belong to a printing company, or the press can be part of the service bureau.

Digital Variable Data Printing—With variable data printing, each printed piece is personalized with text and/or images of interest to a specific recipient. The advantage of digital variable data printing is that the response rate is greater than that of non-personalized correspondence.

When selling digital variable data printing, the cost per piece is the same regardless of the length of run. Therefore, successful sale of this technology to print buyers is based on the understanding that the value is in the cost per response as opposed to cost per piece. The printer becomes a communications services provider rather than just a print provider.

The workflow in this segment differs from a traditional printing workflow. It often includes operations such as Internet submission, file transfer, document creation and management, variable data merging and filtering, document preflighting, proofing, video analysis, postal coding and addressing, postage imprinting, and binding and finishing. One goal of digital variable data printing is to provide the look and quality of offset lithography. Some systems have come close and others have reached this goal.

Digital variable data printing is the fastest-growing segment of the printing industry and has demonstrated the effectiveness of one-to-one marketing. The most important facet of variable data printing is setting up databases of information that allow true personalization. A printer "going digital" has to increase hiring in data processing because a different skill set is needed than that of a designer or electronic prepress technician. For sophisticated variable data jobs, the printer must invest a lot of time and money in front-end technology and personalization systems.

Print Broker—Print brokers connect printing companies and customers. They are independent and not directly employed by the companies they represent; they serve as sales representatives to printing service providers and they bring in work without costing the service provider a commission. Print brokers are compensated by marking up the cost of the printing job.

Print brokers are particularly valuable to those who need printing but know little about how to specify and buy it. Print brokers coordinate all aspects of a printing job beginning with art and copy preparation and extending through prepress, press, and finishing operations. While some have basic copy preparation departments within the brokerage, brokers primarily outsource most facets of a printing job. Some brokers specialize in planning specific printed items while others are generalists and handle many types of jobs. They are typically knowledgeable in most facets of printing and know how to provide specifications. Print brokers integrate the print manufacturing process by coordinating all manufacturing services needed for the job.

Products of the Graphic Communication Industry

The various industry segments produce the following:

Books include dictionaries, textbooks, hardbacks and paperbacks, workbooks, graphic novels, yearbooks, Bibles and religious texts, and others.

Periodicals include magazines, journals, newsletters, newspapers, reprints, preprints, digests, and comics.

Catalogs include business, industrial, consumer, distributor, salesperson, dealer, retail, specialty, and mail order.

Direct mail includes letters, postcards, reply cards, self-mailers, notices, packets, coupons, inserts, and flyers.

Directories include membership, telephone, parts lists, Yellow Pages, internal, professional, price lists, schedules, attendee lists, and governmental.

Financial and legal include policies, quarterlies, 10-Ks, checks, legal statutes, lottery and gaming tickets, annual reports, prospectuses, investor information, and notices.

Packaging includes CD covers, book covers, test packages, record covers, labels, tags, signage, displays, boxes, flexible containers, cans, and cartons.

Technical documents include general manuals, instructions, parts manuals, user manuals, maintenance, industrial product bulletins, guides, repair notes, and applications.

Promotion (advertising) includes flyers, data sheets, ads, ad inserts, brochures, booklets, folders, counter cards, and posters.

Corporate products include reports, presentations, documentation, notices (posters), newsletters, directories, organizational charts, invitations, and flyers.

Miscellaneous products include greeting cards, diplomas, reports, proposals, certificates, business cards, stationery, menus, and wrapping paper.

The growing markets of the printing industry are custom printing, variable data printing, just-in-time printing and publishing, mass-customization of media, on-demand, one-to-one communications, market of one, and targeted marketing. These growing markets rely on some of the latest computerized and digital technologies available for producing printed products and are characterized by short-run color and personalized printing produced rapidly.

6

Paper, Ink, and Toner

Substrates for Printing

A substrate is the material on which an image is ultimately produced and examples include paper, board, plastics, foils, vinyl, and others. Paper is available in various weights: heavy, normal, and light. Paper and board can be uncoated or coated. Board can have a smooth liner to it or it can be corrugated. Paper and board can be made from original pulp or recycled pulp. Paper also can be synthetic and digital, wherein the image can be electronically changed.

Inks and Toners for Printing

Inks and toners have to be specially formulated to meet the printing requirements of the substrate and the printing process being used, and each process has specific ink or toner requirements unique to that process. Some of the variables in ink are viscosity, tack (stickiness), color strength, size of ground pigment particles, transparency or opacity, and more. Inks can be either solvent- or oil-based or water-based. Different processes and substrates allow for one or the other. There are also specialty inks such as metallics that give the feel of a metal surface to the printing. Toner can be wet or dry. Different printing processes require one or the other, and the compatibility of toner and substrate sometimes depends on the nature of the toner and the structure of the substrate. For example, dry toner particles sometimes scratch easily if not properly cured on high gloss or coated paper.

Paper

Paper is the most expensive disposable commodity used in printing, and that is why books cost so much. Paper accounts for between 30 and 50 percent of the total cost of most printed products. For example, if a manufacturer prints an advertising campaign involving slick brochures and spends a million dollars for it, it is conceivable that $500,000 went to cover the cost of paper.

This raises the question: Why is paper so expensive and is anything being done to reduce the cost?

For the most part the public is not aware that there has been a paper shortage in the United States for more than thirty years. When commodities of any nature are in short supply, their prices increase. The national paper shortage dates back to the early 1970s with the founding of the Environmental Protection Agency (EPA) and the enactment of the Occupational Safety and Health Act (OSHA) and the resulting federal agency. What these two agencies did was take aim at most American industries in identifying those that contributed most to the pollution of the environment and to endangering the safety of employees. The paper manufacturing industry was targeted as a major violator in both instances.

The paper manufacturing process involves the use of a large quantity of water. Indeed, when paper is in its forming stages, it is made of approximately 95 percent water. As the water is drained off through the raw pulp material used to make paper on a fast moving machine (Fourdrinier machine) as long as a football field or more, the wastewater goes into rivers and streams. That's why most paper mills are located near large bodies of water. This water picks up contaminants within the pulp and has resulted in the pollution of rivers and streams. At least this is the way it used to be—this is not the case today. Since restrictions imposed and fines levied by the EPA, steps have been taken by paper mills to purify (waste) water before returning it to the bodies of water from which it came.

Additionally, papermaking machines are huge machines costing millions of dollars that take as long as five years to build. They are not things easily replaced. OSHA discovered that many of the papermaking machines used in the early 1970s were built in the early part of the twentieth century—between 1910 and 1920. Hence, they lacked the safety guards and warning signals that more

Fourdrinier papermaking machine
(Océ)

current machinery in other industries were employing. Severe accidents occurred in paper mills as a result.

The combined actions of the EPA and OSHA resulted in citations and major fines if environmental and safety issues were not corrected in a timely fashion. These were enormously expensive corrections that many paper mills simply could not afford and, therefore, a large percent of the nation's papermaking capacity was shut down by the mid-1970s. This caused a grave paper shortage in the years that followed; paper was rationed to printers in the same way that gasoline was rationed earlier that decade. Printers only received a percentage of the paper that they had received in previous years. Paper prices skyrocketed, and American printers were forced to acquire paper from Canada and even from overseas—all of which further inflated the cost of paper.

By the start of the twenty-first century, the paper industry has not recovered from the loss of manufacturing capacity of the 1970s. There was more papermaking capacity in the United States in the early 1970s than there is today.

Are we moving to a paperless society?

The 1970s was an interesting time in the United States. It was the decade of environmental awareness and employee protection. Brought on largely by the publicity surrounding the work of the EPA and OSHA, the American public at large became aware of and outspoken on matters of environmental protection, health, and safety. The issue even arose about the destruction of forestland and

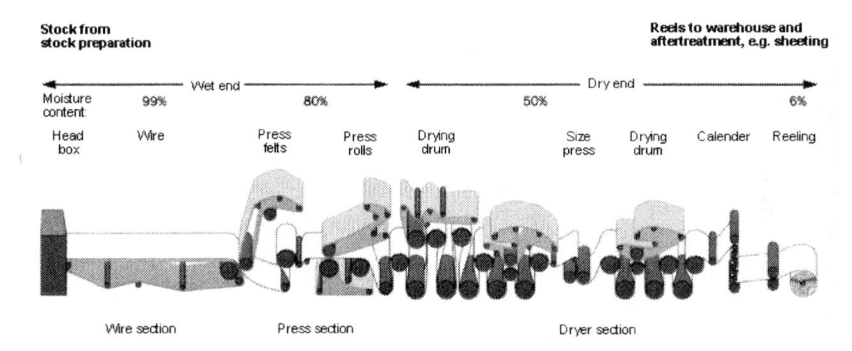

Paper starts out as 95% moisture and ends up at about 6% moisture. (ForestSweden)

the consumption of trees for making paper. The reality of the situation was that the paper mills were not simply removing trees but also had an aggressive replant program where, for example, two new trees would be planted for each mature tree removed. However, this was not a convincing enough argument for the general public because the result was, for the most part, invisible considering the length of time it takes for trees to grow.

Another interesting development in the 1970s was the birth and growth of the office computer industry with all of its promise of technologies that could remove the need for the large quantity of paper consumed on a daily basis by business and industry. Environmentalists, who promoted the concept of the "paperless society," embraced this promise. The question was raised: Is it possible to replace the nation's reliance on paper with forms of electronic communication that would resolve the issues of river and stream pollution, employee safety, and the depletion of forestlands? The decades following the 1970s has proven that paper is a commodity necessary for communication, and the fact is that the volume of paper consumed in the United States has never been greater and continues to grow. The products for which paper is used may be changing, but its overall usage has grown. For example, business forms have become, for the most part, electronic. However, the production of cut paper for use in computer printers and office copiers has grown exponentially. There are fewer daily newspapers in the United States but more special interest magazines. There are fewer printed directories but more direct mail advertising.

Some companies heeded the prevailing concerns of the 1970s and began exploring alternatives to conventional paper made from wood fibers and pulp. An early entrant into substitute paper manufacturing was the Dupont Corporation's invention of Tyvek—a synthetic paper-like material resembling real paper but having a base made from plasticizers. Dupont recognized that the nation would be facing a severe shortage of paper and that the advent of computers would likely increase the consumption of paper, not decrease it. The corporation knew that there was a general affinity to the feel, look, and portability of paper, and Tyvek could be manufactured with fibers resembling real paper or with the type of coatings found on real paper.

Synthetic Paper Methods of Production

There are two varieties of synthetic paper; one produced through an extrusion process such as Tyvek and the other produced with a coating blade. The extrusion process allows the plastic material to flow through a screen-like device made up of very small holes that converts the plastic into fibers resembling

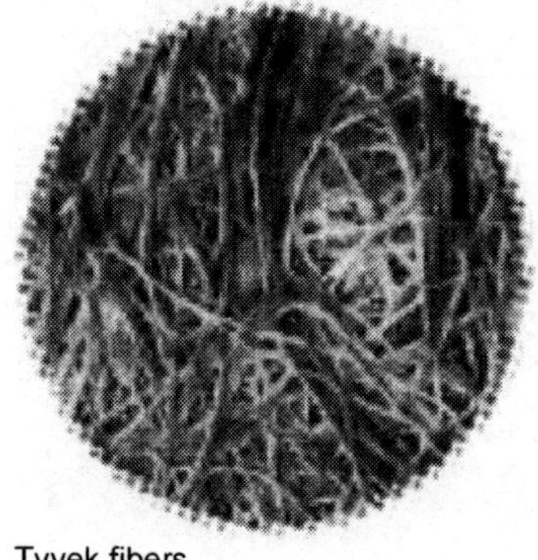

Tyvek fibers
(Dupont)

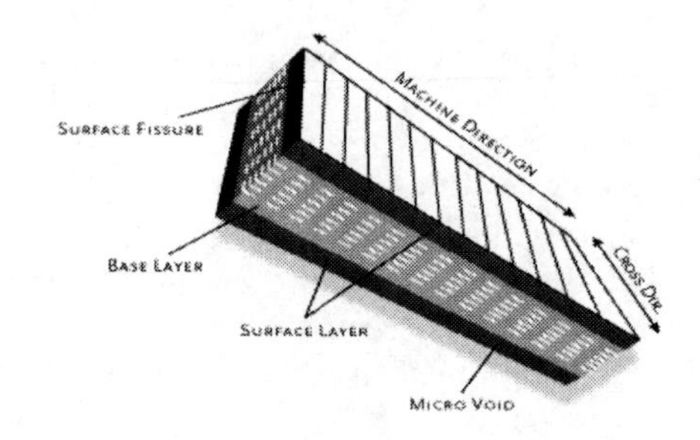

Yupo synthetic paper structure
(Yupo)

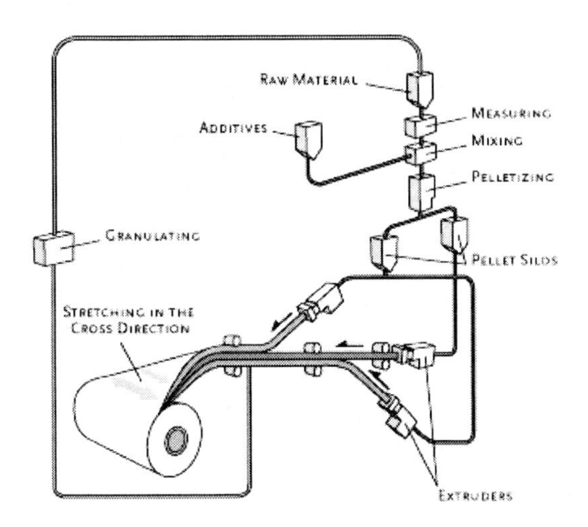

Extrusion process for making synthetic paper.
(Yupo)

paper fibers. The fiber structure can be left visible or covered with a coating. The coating blade process, used by the Yupo Corporation, uses a blade that spreads the plastic into a continuous sheet of varying thickness. The coating blade process does not have the fiber structure of the extrusion process.

Printability of Synthetic Paper

While synthetic paper looks and feels like traditional paper, its printing requirements differ. One important difference that impacts printability is synthetic paper's lack of absorptiveness, also known as "ink holdout."

Whereas traditional paper "dries" to a certain extent by absorption and oxidation, as well as with drying elements such as cobalt and manganese added to the ink, synthetic paper requires more sophisticated drying apparatus. These include ultraviolet (UV), infrared (IR), or electron beam (EB) dryers, which dry the ink instantaneously. There is no time for an oxidation process to take place to enhance ink drying. If the ink is not dried instantaneously, the sheets will stick together.

Additionally, press settings are more critical when printing synthetic paper verses traditional paper. Some of the critical controls on offset lithographic printing presses are plate-to-blanket squeeze, ink film thickness, fountain solution control, form roller setting, roller hardness (durometer), and ink density control.

Inks for synthetic paper are formulated differently, and the handling of inks on printing presses also differs.

Some Reasons for the Development of Synthetic Paper

• Paper shortages

This has already been covered but is worth noting again. Quite simply, there continues to be a paper shortage in the United States due to a greater demand for paper than the capacity to produce it.

• Longevity

Librarians like the promise of synthetic paper because one of the problems they face is the deterioration of paper made from pulp. While there are examples of books that have survived for 500 or more years, a typical publication made of traditional paper deteriorates much sooner due to the acid content of chemicals used in making paper. Significant efforts to reduce or eliminate the acidity of paper have been made in recent years.

• Resistance to moisture and tearing is good for children's books.

Children are particularly hard on books and synthetic paper is more durable and long lasting, as it does not tear and can be washed.

• Size and expense of machines for conventional papermaking

As previously noted, building papermaking machines takes many years and costs millions of dollars. Paper mills are resistant to making the capital investment needed to increase papermaking capacity.

• Ability to be recycled

Contrary to what one might think, synthetic paper does not create an environmental problem when recycled. In fact, recycled synthetic paper provides a number of advantages. For one thing, the synthetic recycling process is easier than that for traditional paper. Traditional paper recycling still requires equipment similar to that used to create it. However, the entire process of creating synthetic paper uses more compact and less expensive equipment.

Furthermore, there is no issue of fiber length as related to paper quality. When recycling traditional paper, the quality of the resulting product is reduced with each recycling. This is because paper strength, an important quality, is for the most part based on fiber length. The longer the fiber, the stronger is the paper. However, with every recycling, the fiber is shortened and, therefore, paper strength is diminished. When plastic is melted down the resulting plastic has the same quality as the original plastic. Hence, we are not dealing with reduction in fiber length as in the case of Tyvek or the overall strength built in the paper caliper as with the Yupo product.

Unlike recycled traditional paper, all components of synthetic paper go into its recycled product. There is no draining off of water and extraneous pulp.

People Like the Look and Feel of Paper

Perhaps the biggest surprise to the proponents of "the paperless society" concept is the reluctance of people to give up paper. Paper has portability as compared to laptop computers or electronic books. Paper is easy to read on busses and trains, in bed, and at the beach.

Paper has high resolution. Consider the crispness and sharpness of pictures on slick advertising pieces or in magazines. The pictures take on a photographic quality.

People are used to the look and feel of paper, and they are reluctant to give up things they are used to that serve a meaningful purpose. Studies have shown that people prefer printed magazines, for example, to online magazines. People are not generally opposed to paper "junk mail" received in their mailboxes at home. It is easy to discard and also easy to save for future reference. However, people are extremely resistant to electronic junk mail or "spam" received on the Internet.

Paper publications are easier to scan from front to back or find things in the middle than their electronic counterparts. One can often find things faster in a printed publication.

Paper is easier on the eyes. Once can read a book for hours at a time without stopping. However, reading a computer monitor creates a great deal of eye fatigue.

Paper products support electronic communication. Nearly every Web site promoting products for sale has some sort of paper counterpart that it sends to present and prospective customers. A good example of this is www.amazon.com. Whereas Amazon is the largest seller of paper books online, it also produces millions of printed catalogs to help sell its online products.

People are often willing to pay for paper products that are printed but resist paying for electronic versions. Printed publications are often perceived to be more credible than their electronic counterparts. For example, there is typically no hesitation in paying for newspapers or consumer magazines, but there is a great deal of resistance to paying for online versions. People expect most of the information contained on Web sites to be free.

Myth Verses Reality

So, with what has been noted so far in this book, are we moving to a paperless society? The myth is that we are, but the reality is that we are moving toward longer lasting and more durable paper. Paper consumption continues to grow.

Traditional Paper vs. Electronic Paper

Electronic paper has the look and feel of traditional paper made from wood pulp or of synthetic paper made from plastic. However, the paper surface performs as a computer monitor does. In other words, you can change the image without the need for additional paper. Think of the paper savings!

Imagine reading, for example, a classic novel such as *Gone With the Wind.* When done, you may want to read *From Here to Eternity.* You take the *Gone With the Wind* version, plug it into a device that activates the electronic paper, and the manuscript is replaced with *From Here to Eternity.* This could also be done with your daily newspaper or weekly news magazine. One practical application of this is changing the image on billboards or other forms of outdoor advertising requiring large image displays. Instead of the tedious and time-consuming task of replacing billboards, they can be replaced by a click of a computer mouse.

The following table provides a comparison of the advantages of traditional paper verses electronic paper.

Traditional Paper	Electronic Paper
Durability	Can surf the Web
Resolution	Masses of information
Feel	Speed of access
Portability	Ease of correction
Review of information	Rapid distribution
Can write on it	
What we are used to	

Electronic Paper/Ink

Two organizations in the forefront of the development of electronic paper were MIT's Media Laboratory and Xerox' Palo Alto Research Center (PARC). While the products of the two organizations are not the same, there are similarities. For example, both are based on spherical geometry and the presence of bi-colored microcapsules built into the paper. The microcapsules are manipulated by electrostatics whereby the colored microcapsules can be brought to or removed from the paper's surface.

E-paper
(Akihabara)

In other words, the microcapsules respond to an electric field or matrix of electrodes built into the paper that serves two different states as an "on-off" switch for each microcapsule. "On" could mean "show color" and "off" could mean "don't show color." Once the device is turned off, the image remains intact. When turned on again, the user can change the image.

The following summarizes the PARC and MIT systems.

PARC's Electronic Paper

- The paper contains thousands of tiny half black, half white balls (microcapsules).
- The microcapsules are approximately 100 microns in size.
- Each ball is floated in oil and rotates in response to an electrical field.
- Electrodes, or switches, tell balls when to rotate.
- A pattern of voltages is applied to the paper that rotates the microcapsules to create text and pictures.
- A new voltage pattern creates a new image.

MIT's E-ink

- The paper contains millions of microcapsules.
- Each microcapsule contains white particles in dark dye.
- When an electric field is applied, the white particles move to one end of a microcapsule.
- The particles become visible and the surface appears white at that spot.
- An opposite electric field pulls the particles to the other end of the microcapsules.
- The particles hidden by the dye make the surface appear dark at the spot.
- By applying an electric field so that only some of the white particles float to the top of the microcapsule, various shades of darkness and light can be created as is needed in producing a picture with various tones.

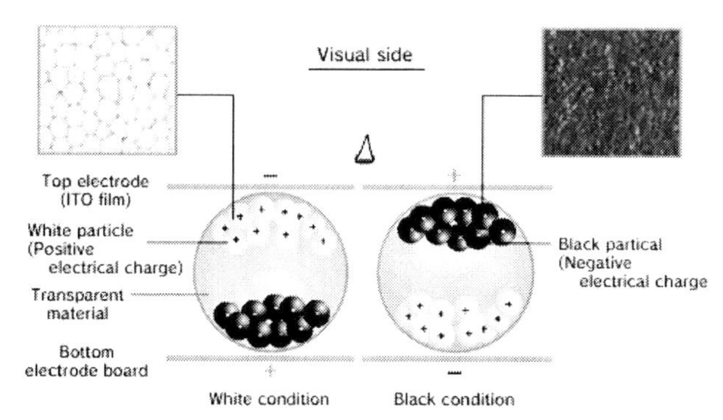

Electronic paper based on the MIT model.
(Midori Mark)

Radio Frequency Identification (RFID) technology

Related to electronic paper is the concept of Radio Frequency Identification (RFID). This is a technology used to change an image on a surface from a remote site using wireless technology that is not influenced by obstructions. RFID has great promise in the fields of outdoor advertising, counter displays in stores, and packaging.

RFID is the use of small devices that can be identified using radio waves or a nearby frequency with few problems of obscuration or orientation. The technology was first used more than fifty years ago on aircraft, and transport remains the largest application today (such as for car access clickers to smart cards for ticketing and windshield tags for road tolling).

One example of applications being explored is electronic billboards that are equipped to profile commuters as they drive by and then instantly personalize freeway ads based on the wealth and habits of those drivers. For example, if a freeway is packed with country music listeners, the billboards might make a pitch for entertainment venues featuring country music.

A basic RFID system consists of three components:
An antenna or coil
A transceiver (with decoder)
A transponder (RF tag) electronically programmed with unique information

RFID technology made its entry into the field of packaging several years ago. The requirement in packaging is usually for a very thin, low-cost tag that allows changing prices and other relevant information on packages when appropriate. This could include marketing information and nutrition information. Innovative product benefits, such as today's talking label for the blind, will play a major part in the widespread use of RFID in packaging.

Today the RFID device used in packaging is usually a complete tag, protected against the elements. However, some such devices can be buried in the plastic or wood of a crate. Other tiny versions can be buried in the paper of packaging, such as a security stripe in a banknote. Intermediate types can be buried in corrugated paper of packaging or in cardboard. Others are best fitted on the inside surface of a pack or in seams.

RFID can be used to alter package pricing on the supermarket shelf. (Packageworld)

In sum, electronic paper makes printing economical because a sheet can be reused. Present resolution is up to 600 dpi, with some of the immediate applications being focused on outdoor advertising and packaging. Resolution will increase as the technology develops. The future will show developments of electronic paper for direct mail advertising and consumer publications including catalogs, newspapers, and magazines.

Ink

Whereas paper is the most expensive disposable commodity used in printing, ink is the second most expensive. As the paper industry improves technology and operations to deal with the environment and diminishing supplies, the ink industry, too, is dealing with similar issues. Ink does not face a shortage problem, but it does face issues of quality and of the environment.

When faced with the question "What do you like least about newspapers?" most people say, "The ink rubs off on my hands and clothing." This happens because in newspaper printing, where it is imperative for the publisher to keep the cost of producing the product as low as possible, the ink used is not as expensive as or of the same quality as inks used for slick and glossy magazines and advertising pieces. By eliminating chemical dryers from the inks and by not placing expensive drying mechanisms on newspaper presses, the publisher keeps production costs down. On newspapers or other publications composed of newsprint, the ink never really dries but simply is absorbed into the paper, where it sets. The oils and pigments in the ink are what rub off. Some oxidation occurs, but the ink never really dries. One way of rationalizing this is to point out that for the most part newspapers have a life of one day, after which they are discarded. This is not true of magazines. However, there is hope for the consumer as research and development continues in the formulation of water-based inks to replace oil-based inks. Some newspapers have already implemented the use of water-based ink and have solved the ink rub-off problem for their readers. For example, if you purchase a *San Francisco Chronicle* and rub your hands over the printing, the ink will not rub off because that newspaper uses water-based ink. Then, you may ask, why don't all newspapers convert to water-based ink? The answer is that making this conversion requires expensive press retrofitting or converting to flexographic presses from the offset lithographic presses commonly used to print newspapers. These printing processes and others are described in detail in this book. Flexography lends itself more to water-based ink than does offset lithography. New presses typically cost millions of dollars and not all newspaper publishers can afford to make such investments.

With the solution of the ink rub-off problem comes a way of addressing environmental concerns of the ink manufacturing industry.

Solvent-based Ink Versus Water-based Ink

A world without volatile organic compounds (VOCs) is a world without pollution. Oil- or solvent-based printing ink contains VOCs; water-based ink does not. Anyone concerned with the air they breathe, or the world their children will inherit, must understand and accept that the greenhouse effect is real. We have to work hard to create a sustainable planet. Substituting water for solvents would be a major step toward improving the environment and the health and safety of workers.

Water-based ink is highly promising for publication printing requiring marginal quality image reproduction on non-coated paper. It does not provide the gloss or finish of solvent-based inks or the ink holdout on the top of the sheet. Therefore, water-based ink poses some problems for package printing, where the highest print quality is required. Additionally, much package printing is produced on non-paper substrates such as foil and cellophane. Such substrates do not lend themselves well to water-based ink.

The positive aspects of water-based inks are that they work well for paper on which inks are absorbed and they are rub-resistant when dry. They are popular with the flexographic process on non-coated paper and they eliminate smearing. Water-based inks are environmentally friendly because they do not emit volatile chemical vapors into the environment. Therefore, no incinerators are necessary [incinerators convert VOCs into carbon dioxide, which is not good for the environment].

Water-based ink requires fewer procedures in dealing with chemicals and comes under fewer state and federal regulations. The printer is also less concerned with issues of fountain solution pH and the viscosity of the ink and encounters fewer ink film thickness problems.

The Upside of Water-based Ink

- Works well for paper on which inks are absorbed
- Does not rub off
- Popular with flexographic process
- Eliminates smearing on newspapers
- Environmentally friendly (no emissions of volatile chemical vapors)
- No incinerators
- Fewer procedures in dealing with chemicals
- Fewer problems with state and federal regulations
- Fewer ink film thickness problems and pH and viscosity concerns

However, water-based ink does pose printing problems that have slowed its adoption. The use of water-based ink requires the need for specialized technology or retrofitting of present presses that use solvent-based ink. As already noted, it is ineffective for competitive packaging, and the adhesion to smooth surfaces is less than that of solvent-based ink. There is less gloss and drying time is longer, which necessitates slower printing press speeds.

The Downside of Water-based Ink

- Need specialized technology or retrofitting of present presses
- Ineffective for competitive packaging
- Adhesion is less
- Gloss is less
- Drying time is longer
- Printing speed is slower

Desktop Ink/Toner and Paper Issues

While to this point emphasis has been placed on commercial printing, the desktop publishing revolution of the 1990s has resulted in an enormous use of paper, ink, and toner by consumers and content creators. Content creators are those who prepare copy for reproduction. While this used to be the domain of the commercial printers, in more recent years it has become the work of service bureaus, design studios, and related organizations that have taken on much of the prepress work in preparing copy for printing. In many cases the author, publisher, or printing customer have taken on this role.

In deciding the best mix of equipment and consumables to use, there are a number of important considerations. For example, will color be used? If black is the only color that will be used, it might be best to acquire a laser printer. If color is going to be used, an inkjet printer may be better. Laser printers are more expensive; however, the consumable toner cartridges typically last much longer than the ink cartridges used in inkjet printers. The cost of a black toner cartridge will likely be less expensive than a full-color cartridge. Manufacturers' profits are built into the consumables, such as ink and toner, more so than they are built into the equipment. That is why the printers themselves are typically reasonably priced. One major manufacturer of inkjet printers claims that approximately 55 percent of the corporation's profits come from selling inkjet cartridges.

Some Differences Between Laser Printer Toner and Inkjet Ink

Laser toner is baked onto the substrate and does not smudge. It provides high ink density and produces type that is very well formed. Toner particles are not well suited for very smooth or high-gloss paper. On the other hand, inkjet ink smudges easily unless a protective coating is placed over it. This is because inkjet ink is water-soluble and even small exposure to moisture enhances smudging. Inkjet ink does, however, provide high quality color on the appropriate paper, and some inkjet printers allow printing up to six colors on standard paper.

It is highly recommended that the inkjet ink used is produced or recommended by the printer manufacturer. While there are lower-cost generic inks available, as well as refilling apparatus, these alternatives have been known to cause problems. Also, printer warrantees may be voided when using these alternatives. The printers should be used at least once a week to avoid clogging of the inkjet nozzles, and cartridges should be removed and the nozzles covered up if they are not going to be used for longer than a week. If the nozzles do clog, the printer should be run through its nozzle cleaning steps. If this does not resolve the problem, using a Q-tip and a tiny drop of solvent such as Windex may solve the problem. However, using too much solvent can ruin the cartridge.

The paper used for desktop printers should be appropriate for the desired quality. Use standard bond paper for text and line illustrations only. However, use photo paper for printing full-color pictures or for printing tone gradations that must be smooth. Photo paper or photo-quality paper is of higher quality but is more costly than bond paper.

There are two printer styles: paper feed tray and top loading. The paper feed tray style houses the paper cassette in the front of the printer. When printing, the paper is lifted and placed to the rear of the printer. The paper is then printed, turned 180 degrees, and delivered. This feeding mechanism may not be the best if printing on heavy paper because it has to bend 180 degrees. Heavyweight

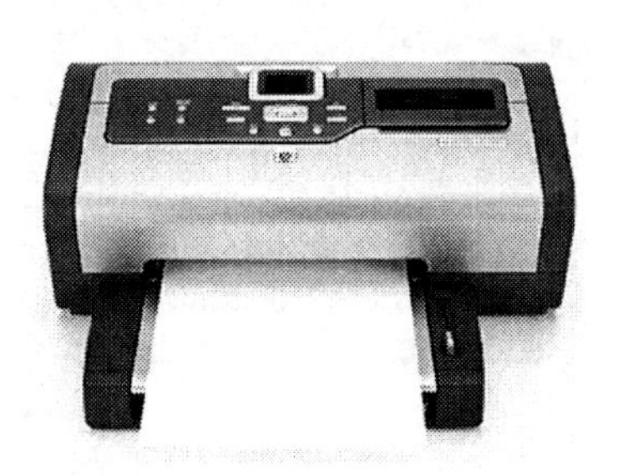

Tray feed printer
(Hewlett Packard)

Top loading printer
(Epson)

paper can resist such bending. Top loading feeders are gravity fed. The paper moves into the printer and is printed and delivered with minimal bending. Gravity feeding and minimal bending makes such printers better suited for heavier papers.

Other paper considerations are room temperature and the relative humidity (RH) of the room in which the paper is used. If the temperature and RH in the room differ from the temperature and RH of the paper, paper curl or waviness could occur, which in turn can cause paper-feeding problems in the printer.

7

Prepress

Prepress can be traditional, electronic, or a combination of the two. Traditional prepress involves a lot of handwork and special skills, while electronic prepress relies on computers to replace human craftsmanship. Prepress encompasses operations such as art and copy preparation, typesetting and imagesetting, graphic arts photography and scanning, image assembly (which used to be called stripping), proofing, and platemaking.

Art and Copy Preparation

Art and copy preparation usually involves the preparation of any combination of line copy, continuous tone copy, or full-color copy. Line copy is comprised of solid images with no tonal variation, such as a pen and ink drawing. Type in and of itself is line copy, as it involves no tonal gradations. Continuous tone copy is typically art containing continuous shades of gray from light gray to dark gray. A black-and-white photograph is an example. Full-color copy is artwork in which all colors of the visible spectrum appear. Examples are color photographs (traditional and digital), color transparencies (such as 35 mm slides), and paintings.

Electronic art & copy preparations studio. (Printing Services)

Preparing the three categories of copy for printing requires either graphic arts photography/scanning or digitization on a computer. Photographing or scanning line art is a simple matter and traditionally results in a film negative or positive used to expose a printing plate. Producing digital line copy on a computer is now commonplace and is also a simple matter due to easy-to-use software available for this. However, preparing digital continuous tone copy and full-color copy for printing is more complex. Traditional methods of doing this have been around for more than one hundred years and the processes involved are generic for preparing printing, regardless of the end-product being printed.

Art and copy preparation today involves the use of application software designed to perform many functions ranging from producing original artwork for cross-media purposes, such as for print and the Web, to sophisticated image manipulation incorporating sizing, retouching, and cloning. There are numerous applications available with new versions and upgrades made available frequently. The following are some popular graphic arts applications and their capabilities.

Adobe Acrobat Reader
Adobe Acrobat Reader is software that permits viewing and printing PDF files (Portable Document Format files) on all major computer platforms. It also permits filling in and submitting Adobe PDF forms online. An expanded version of Acrobat Reader for Windows offers additional functionality, including support for the visually impaired and the ability to search collections of Adobe PDF files.

Adobe eBook Reader
The Adobe Acrobat eBook Reader software enables reading high-fidelity eBooks on a notebook or desktop computer. No special hardware is needed. This reader software displays eBooks with the pictures, graphics, and rich fonts typically found in printed books. It has an intuitive interface.

Adobe GoLive
Adobe GoLive allows designing, building, and managing dynamic content for the Web and wireless devices. It includes the Adobe Web Workgroup Server, which offers asset management features such as version control that enhances workgroup collaboration and allows delivery of personalized multimedia content for a worldwide audience.

Adobe Illustrator
Adobe Illustrator CS is the industry standard vector graphics creation software. Illustrator CS provides options for Web designers, graphic design professionals, and business graphics users. It includes built-in Web graphics creation and optimization tools; unlimited transparency capabilities; abundant new creativity and productivity features; and integration with Adobe's family of professional graphics programs for print, Web, and dynamic media.

Adobe InDesign
Adobe InDesign is the Adobe page layout and design tool for professional publishing that allows combining pictures, text, typography, writing, editing, and printing in one application.

Adobe PDF Writer
Adobe PDF Writer allows the creation of Adobe Portable Document Format (PDF) documents. PDF documents are viewable on all major platforms with the use of the Adobe Acrobat Reader.

Adobe Photoshop
Adobe Photoshop software, the professional image-editing standard, is a tool for image manipulation such as sizing, electronic retouching, cloning, and more.

Macromedia Dreamweaver

Dreamweaver is an HTML editor for visually designing and managing Web sites and pages. It provides a means to either code HTML or work in a visual editing environment. Dreamweaver has many tools for creating and managing Web sites. Its visual editing features allow you to add design and functionality to pages without writing a line of code.

Macromedia Fireworks

Fireworks creates the smallest, highest-quality JPEG and GIF graphics in just a few steps. It is a solution for creating and producing Web graphics. Fireworks eliminates having to recreate Web graphics from scratch after editing. It generates JavaScript, thus making rollovers easy to create, and its optimization features shrink the file size of Web graphics without sacrificing quality.

Media Player

Windows Media Player enables a full range of digital media activities, including playback of CD audio, streaming and downloaded audio and video, jukebox capabilities for CD recording, media management, Internet radio, and integration to support portable music devices.

Microsoft Office

Microsoft Office is a package of software for day-to-day office functions. All Office packages include Word for word processing, Excel for spreadsheets, Outlook Express for e-mail, Access to create computer databases, and PowerPoint for presentations and slide shows.

QuarkXpress

QuarkXpress is an integrated page layout and publishing program that allows combining pictures, text, typography, writing, editing, and printing in one application.

The Halftone

Preparing continuous tone copy for printing, for example, requires converting the copy into a halftone. A halftone is the breaking down of the various gray tones of continuous tone copy into solid dots of various sizes where the smaller dots represent the highlight or lighter areas, the middle sized dots represent the mid-range tones, and the larger dots represent the shadow areas or dark tones. Though different in size, the dots are equidistant from center to center. In other words, the frequency or number of dots per linear inch is the same. When viewed, an optical illusion occurs where the viewer cannot see the individual solid dots because the eye blends the solid dots with the white paper around the dots and the image

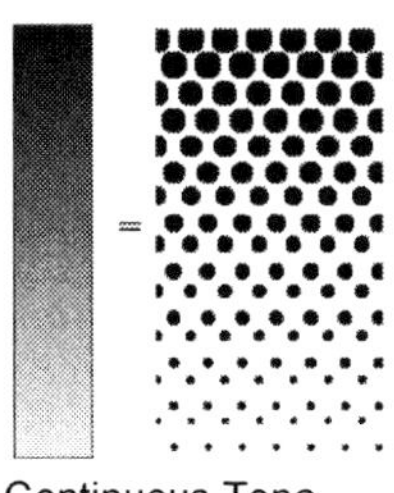

Continuous Tone and Halftone (Xaraxone)

appears in the shades of gray of the original continuous tone copy. Halftones are produced either photomechanically on large graphic arts cameras or on scanners. They can also be produced on computers. When done in the traditional manner, the continuous tone copy is photographically reflected through a halftone screen onto film. The halftone screen is a piece of film that has a microscopic checkerboard pattern on it. The pattern breaks the various tones of the continuous tone copy into dots of different sizes depending on whether the tone is light, mid range, or dark.

A variation of a halftone is a screen tint. A screen tint is a printed area that is not a solid ink, but rather consists of dots of a particular color ink at a particular percentage in size—such as 10, 20, 30 percent, etc. The higher the percentage the darker the tint; the lower the percentage the lighter the tint. This allows the creation of many different color values and many different effects from just one color.

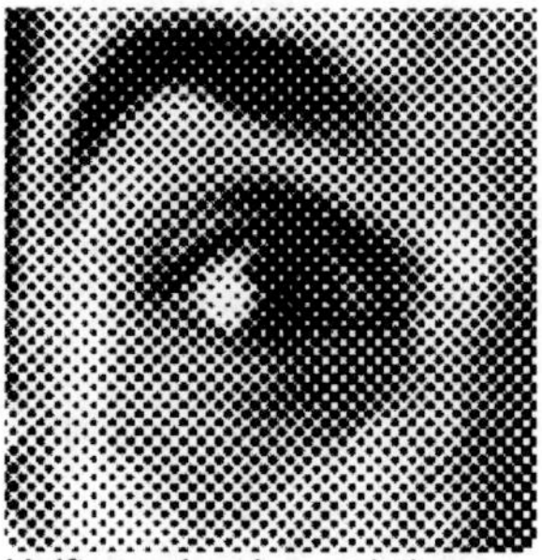

Halftone showing variation in highlight to middletone to shadow dot sizes. (Ollé)

Another variation of the halftone is the duotone. The duotone is typically a reproduction of continuous tone copy using only two colors. However, both colors are halftones overprinted to create the desired visual result. There are also two versions of the duotone: "true duotones" involving the overprinting of two halftones made from one piece of continuous tone copy and the "duotone effect." The "duotone effect" involves overprinting one halftone made from continuous tone copy with a screen tint of another color. Hence, the halftone is printed in one color and then overprinted with a screen tint in another color.

The reason continuous tone copy and color copy must be broken down into solid dots is that the typical printing press cannot print continuous tone shades of colors but only solids—this is true even if the solid is the size of a nearly microscopic dot. Halftones and screen tints can be printed directly on the surface of a substrate or they can be overprinted on top of previously printed images or halftones.

Color Separations

The printing of full-color copy requires that the copy be broken down into color separations. Typically, four-color separations are needed to reproduce all colors in the visible spectrum. However, sometimes five, six, or more colors are used to increase the color gamut (spectrum range) in which detail can appear. Each color separation is a halftone produced either on a traditional graphic arts camera (or scanner) or on a computer. Four-color printing, also called process printing, has been the most common way of printing full-color copy for over a century. Five-, six-, and seven-color printing, sometimes called "HiFi" (as in high fidelity) printing has been used since the 1970s.

Printing plates are made from the separations, one for each color. (International Paper)

The four halftones used in four-color process printing represent the printing ink colors of yellow, magenta, cyan, and black. These inks, once they reach the paper, are so thin that they are transparent and act as filters that transmit or reflect light. By overprinting yellow, magenta, and cyan transparent halftone dots at prescribed angles, all colors of the full-color copy can be recreated. For example, the areas where yellow and magenta overlap produce red; magenta and cyan produce blue; and yellow and cyan produce green. So, in the end, the final print is composed of yellow, magenta, cyan, red, blue, and green. Black is added to enhance the contrast and detail of the final reproduction.

This is a simplistic explanation of a process that is more complex and involves a subtractive color theory where color is taken away to produce color as opposed to being added (additive color theory) to create color. According to the subtractive theory, all colors of the visible spectrum—but particularly red, blue, and green—are already built into the white paper on which full-color copy is reproduced. In physics, an overlap of red, blue, and green light produces white. In printing, when the thin, transparent ink film "filters" of yellow, magenta, and cyan are placed on the white paper, these ink filters absorb or reflect the red, blue, and green present in the white paper.

Yellow, magenta, and cyan are primary printing colors, and red, blue, and green are secondary printing colors. In other words, red, blue, and green are produced by partially overlapping yellow, magenta, and cyan.

In comparison, producing an oil painting, for example, uses the additive color theory, where opaque (rather than transparent) paints are mixed to create

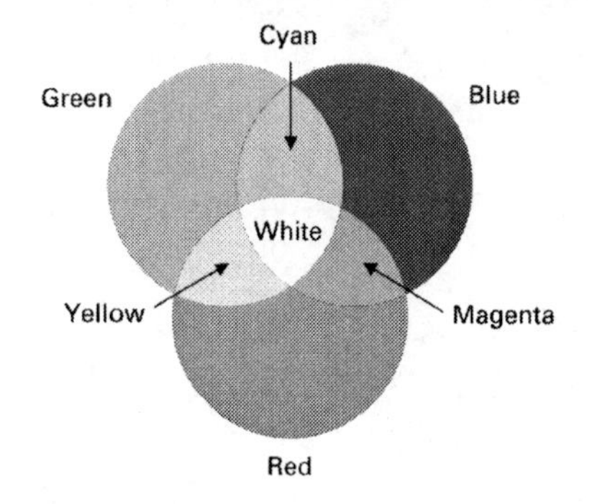

The colors produced when overlapping colors from light.
(Océ "Digital Printing")

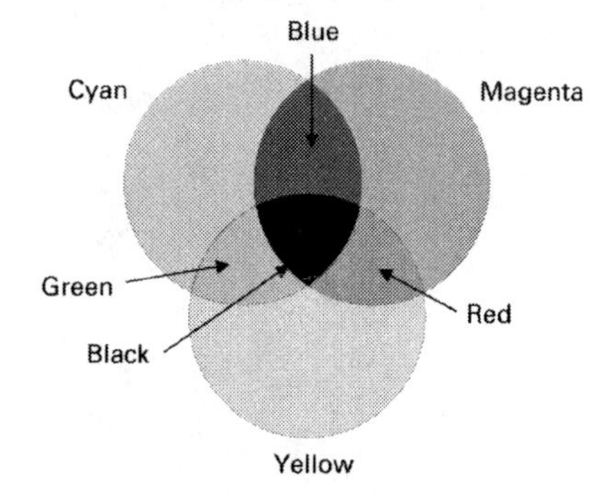

The colors produced when overlapping colors from ink.
(Océ "Digital Printing")

additional colors.

Another version of halftone screening and full-color printing is stochastic or frequency modulated screening. This is a screening technique that does not use fixed screen angles or frequencies. This method produces very small, identical size dots with random appearing spaces between them and requires equipment that can repeatedly produce 14 to 20 micron spots (1 micron = 1 millionth of a meter). This is an electronic digital process and cannot be done using conventional color separation techniques. Its advantage, some claim, is that it results in smooth tones that more closely resemble the tones of full-color copy. It is a process that is often used along with HiFi printing in producing some of the finest full-color printing possible.

The final phases of prepress typically involve image assembly, proofing, and platemaking.

Image Assembly

Image assembly involves assembling pages for printing in the order and position that they will appear on a printed sheet. Once the sheets are printed, folded, and trimmed, the page sequence is correct—page 2 follows page 1; page 3 follows page 2; page 4 follows page 3; etc. Image assembly can be done manually or electronically.

Traditional Film Image Assembly
(Kodak Polychrome Graphics)

Electronic Image Assembly

Proofing

Proofing involves making a facsimile of what the final printed piece will look like so the client has the opportunity to approve or request changes before the print job is produced on a printing press. There are two categories of proofs: hard proofs and soft proofs. Hard proofs are produced on a substrate. A hard proof is a physical piece that a client can handle and write on. Soft proofs appear on a computer monitor only. The advantage of soft proofs is that they can be transmitted quickly to as many people as necessary. The disadvantage is that they do not yet provide the color accuracy that hard proofs do. There are several different systems for producing hard proofs.

For most high-quality printers, a key part of the workflow has been the contract proof. The proof is a document presented to a client for approval prior to printing. Printers wishing to provide high value to their customers now have an expanding range of proofing options.

Selecting ink colors from proof.
(Kodak Polychrome Graphics)

Contract proofs have changed dramatically. Yet even with developments such as laser scanning, desktop publishing, and computer-to-plate systems—which have improved the quality of the printing process—trends in contract proofing are often driven more by cost than by quality.

Soft Proofs – Soft proofs are images that appear on a monitor as opposed to hard proofs on a substrate such as paper, film, or some other material. Soft proofing detects errors and speeds the proofing and approval process. In soft proofing, the cyan, magenta, yellow and black bitmaps are converted into red, green, and blue for the monitor.

Soft proofing does not necessarily eliminate the need for a hard-copy proof in some cases, but it does allow the correction cycle to occur faster and reduces the number of hard proofs that are needed.

Soft proofing systems are improving and are becoming more popular for comparisons at the press. However, the process demands a lot from the monitor and the lighting environment. All parts of the system have to be thoroughly calibrated. Today, soft proofing is useful in providing press operators with a general idea of what the printed page will look like. One current problem is that soft proofing is not user-friendly to customers or production operators for making changes or corrections or for transporting proofs between operators and customers. There is also the issue of two or more monitors on which the proof is viewed being calibrated differently and, hence, displaying images that appear different in color. However, advances in color management tools, improved electronic file transfer methods, and upgraded computer monitor technology are prompting printers to pursue soft proofing methods.

Calibrating the monitor requires determining its color gamut and adjusting it to match the colors that the printing press will produce. This must be consistent among all parties viewing the proof. Soft-proofing methods can range from PDF files via e-mail or FTP sites, to systems that allow users to review content and layout, to more complex color-managed systems that seek to emulate hard-copy proofs. While users may adopt soft proofing as a workflow tool to speed production, whether they embrace it as a replacement for the hard-copy contract proof depends on the comfort level of printers and print buyers.

Hard Proof (Right) Soft Proof (Left)
(GTI Graphic Technology)

While a large amount of current online soft proofing simply involves posting PDF and JPEG files on password-protected areas of a Web site, many e-commerce vendors and even some makers of prepress-workflow systems have added online soft proofing capabilities to their products. Online soft proofing is especially valuable for projects with tight deadlines. However, the process is not suitable for all situations and particularly not for those involving color-critical projects because calibrating displays can be difficult. But for applications that warrant it—such as repeat jobs, projects involving Pantone or spot colors, or clients who trust the production vendor to produce accurate color—the process can be valuable.

Remote Hard Copy and Remote Proofing – Traditional analog and digital halftone proofing systems have a high acquisition cost, making them inaccessible to many smaller printers and impractical for low-volume remote-proofing applications. Lower-cost inkjet proofing systems can deliver prints with six colors at high print resolution. These printers are powerful proofing tools, capable of output that simulates traditional analog proofs. They have color profiling capabilities and the ability to print on special media.

Remote proofing involves sending files to clients (or vice versa) for output on a proofing device, or soft proofing via a computer monitor. Either or both can be integrated into a printer's workflow depending on client needs and expectations.

Wide Format Digital Hard Proof

The availability of lower-cost alternatives that simulate the output of analog and digital proofing systems lowers the price of proofing and places hard-copy color proofing at more points in the workflow. To ensure consistent print quality between systems, some manufacturers have integrated densitometers that automatically make color measurements allowing adjustments to keep color values within specifications. Densitometers are standard monitoring devices in the printing industry.

Electrophotographic color copiers and printers are also being used as proofing devices. They may be more expensive to acquire than inkjet proofing systems; however, they usually have a lower color-page cost. The lower page cost and a larger paper capacity also make electrophotographic units more suitable than inkjet for printing short-run color printing. Hence, a color copier can be a dedicated proofing system and also handle other graphic and office printing.

A challenge to printers is getting their customers to adopt remote proofing. The long-standing stable accounts that generate high dollar volume are more receptive to remote proofing than newer customers or customers with infrequent print work. Wide acceptance of remote hard copy proofing also depends on the development of inexpensive systems that are easier to implement and maintain.

Not all remote-proofing workflows rely on the transmission of job files to a client for output on a remote proofing system. Some printers install color-calibrated digital proofing systems at customer locations and let them produce their own proofs, which are submitted along with the job files.

Remote proofing in general has presented challenges for printers with critical-color clients, but emerging new technologies and products are addressing this issue. Aiding remote proofing are inkjets with improved color gamut, light-fastness, and stability. In addition, the cost of consumables on these inkjet printers is lower than that for high-end digital proofing systems. Other considerations that may spark widespread adoption of remote proofing include increased availability of low-cost color measurement devices and profiling software.

For color-managed remote proofing to succeed, there has to be a relationship between the designer, the print buyer, and the printer so that all involved agrees on paper, standard ink sets, and all of the components that influence color. One of the most important considerations is the ambient condition. In other words, the lighting under which the proof is viewed must be controlled.

Platemaking

Platemaking involves transferring the assembled images onto an image carrier used on the printing press. Conventional platemaking for some processes requires exposing the images as film negatives or positives onto a printing plate having a light sensitive emulsion. After the exposure is completed, the plates are then developed using processing machines and chemicals. Besides the exposure and developing technique, some processes use etching systems for making printing plates, while others use engraving systems. These conventional systems have been used for generations.

Computer-to-Plate System (Castle Press)

More current platemaking involves what is known as "computer-to-plate," where the assembled images are electronically transmitted to printing plates. Some systems have eliminated chemicals for developing and instead use water. Others require no developing, and images are produced by laser beams via light intensity or thermally via heat intensity. Another version of platemaking is called computer-to-press, where plates are imaged electronically after they have been mounted on the press, or plate cylinders, having no removable plates, are electronically imaged directly from a computer.

Electronic Prepress

The trend in prepress is away from traditional methods and toward electronic methods. Halftone dots are now defined in pixel resolution, whereby each dot is composed of multiple pixels. The issue of dot frequency has given way to spot frequency. With spot frequency, the dot is related to the halftone or screen tint, and the spot is equated with pixels or stochastic screens. The present focus of prepress is on file management, workflow analysis, preflighting, and application software.

Building files that have complete electronic data for printing a job is quite complex. For companies to stay competitive in producing the greatest amount of work in the shortest amount of time accurately, proper workflows—from the start of a printing job to the finish—must be employed. Preflighting is a procedure for making all adjustments and corrections to an electronic file and for ensuring that the file has all of the necessary information needed to print a job flawlessly. Preflighting software is used to achieve this.

Finally, there are numerous application software options for prepress and they change rapidly. Staying abreast of the proper applications to achieve the desired results and of the ongoing upgrades is imperative for a printing company to stay competitive, productive, and efficient.

File Management

Digital prepress involves working with files comprised of all of the elements that make up the image to be printed. Proper management of these files is vital. File management consists of a set of interrelated steps designed to ensure that files can be readily identified, organized, accessed, and maintained. Since there are strong connections between various aspects of file management, planning ahead is necessary to avoid making decisions that limit options later on. It is important to keep the lines of communication open between technical staff and project staff during the planning stage.

File management steps include the following: keeping track (basic file system considerations); image databases and other image management solutions (special software for organizing image files); storage (devices and media); and maintenance (backup, migration, preservation, and security).

File management is important because default file and directory naming schemes are rarely optimal for a specific collection of information. Sound decisions about files and directories can help minimize problems, especially for large collections. The nature of the images being scanned will suggest organizing principles. For example, serials are often divided into volumes and issues. Monographs have page numbers, and manuscript or photograph collections have folder or accession numbers. In most cases, some aspect of these physical organizing principles can be translated into file system organization.

File system recommendations include:
• Using a file-naming scheme that is compatible with whatever operating systems and storage media being use.
• Using standard file extensions for different file types.
• Avoiding overloading directories with too many files.
• Relying on storage management software to manage large collections across multiple disk drives.
• Allowing for large amounts of collection growth.

Organizing metadata is important in file management. Metadata is a description of elements that comprise a document and helps to facilitate search and retrieval. For example, an address label contains a name, address, city, state, and zip code; the metadata is these five elements. Therefore, metadata is the set of associated attributes or descriptions of information objects and gives them meaning, context, and organization. Metadata is embedded in the cataloging of print publications. In the digital realm, additional categories of metadata have emerged to support navigation and file management. Metadata describes other data.

Page Description Language

Page description language is a computer language or software that describes an entire page—including text, graphics, lines, and halftones—as a series of codes. It allows for viewing or output on any device capable of decoding the language.

PostScript is the most popular page description language. Invented by Adobe Systems, PostScript consists of software commands which, when translated through a raster image processor (RIP), form the desired image on an output device such as a laser printer or an imagesetter. PostScript is commonly used for text and line art. In the latter case, it is referred to as encapsulated PostScript (EPS).

PostScript's advantage is that the code is rasterized in an interpreter in the output device, not the computer. Therefore, images area readable or printable regardless of the computer platform being used.

Essentially, PostScript allows for device independence, or the ability to generate virtually identical output on devices made by different manufacturers, so long as they can interpret PostScript commands. The original version of PostScript is known as PostScript Level 1 and was one of the most important elements in the invention of desktop publishing. A second, revised version of PostScript, called PostScript Level 2, includes support for process color output. It provides support for process color output and particularly defined color spaces and algorithms for the data compression of color images.

Raster Image Processor (RIP)—RIPing

A RIP is a microcomputer that uses a laser to create images out of lines of dots (rasters). It translates commands in a page description language, such as PostScript, or in other languages—PDF, TIFF, EPS, JPEG—into a precise image of a page. Everything a RIP produces is comprised of pixels. The RIP then sends the rasterized data to a recording device (imagesetter) for output on paper, film, or plates.

To create a halftone dot, the RIP divides the area to be imaged into halftone cells. Each halftone dot then is created out of the pixels available within each halftone cell. The speed of a RIP generally determines how fast a printer can produce the first copy of a page.

Preflighting

As noted earlier, preflighting is the process of checking an electronic document before it is moved to the printing process. The process makes sure that all of the elements are correct and within the specifications for printing. Since nearly all work submitted to printing companies is in electronic form, files sent electronically from a remote site or delivered on a disk must be verified for consistency, presence of graphics and font files, and proper selection of colors. Those involved in preflighting check the quality of the data file and correct problems.

If a preflight technician discovers an error in a file, the file can be fixed by the printer or returned to the originator for repairs. In either case, costs often go up and deadlines are missed. Preflight checks include CMYK mode, cropping, bleeds, photo sizes, resolution, retouching, masking, gray balance, proper spelling, page order, creek, type size and style, trapping, clipping paths, fewest anchor points for clipping paths, dot gain, flattening of layers, screen lines per inch, and screen angles. A preflight technician will also check to see that files were saved in the proper format and that files reflect the most recent changes.

Imagesetting
Imagesetting is a method of exposing photographic paper or film using raster (also known as bitmap) or vector techniques. To produce these materials used in printing, imagesetting uses electronic files rather than conventional camera

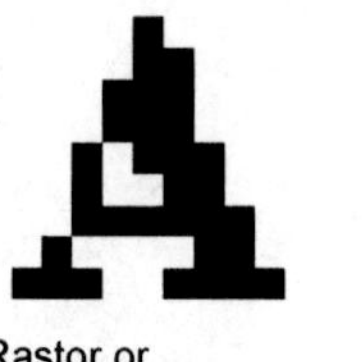
Rastor or
Bitmap type

Vector type

technology. It yields higher quality and less expensive printing. Platesetters are also imagesetters. They are high-resolution imaging devices that use a laser to either burn an image onto a metal plate or to expose photo-sensitive plates. This process is also referred to as computer-to-plate (CTP). It allows producing quality that is better than traditional film-to-metal plates. CTP can cost less and allows for quicker turnaround times, but it requires investment in equipment, training, and changing workflows.

Manufacturers recognize that imagesetter users are under ever-increasing pressure to produce quickly and inexpensively—not only because electronic media are raising customers' expectations regarding throughput times, but also because print runs are becoming shorter.

Some imagesetters use multiple lasers and a 30,000 to 40,000 rpm spinning mirror that traverses inside a drum. The lasers sit next to each other and use the same optical path. Multiple lasers increase speed. To keep costs down, some imagesetters use a single beam that is split with the beams being deflected. A five-faceted spinning disk with a holographic deflection surface is used to direct the laser beam onto the film surface. Each rotation of the deflector produces five scan lines making it up to five times faster than a conventional mirror-based system at the same rotation speed.

Imagesetter RIPs are software-based and run on standard, fast, and easily replaceable workstations.

8

Printing Processes

The Printing Press

The printing press is the basic unit of the printing process. It is a precision instrument, typically the largest and heaviest of all equipment used in printing, but controllable to thousandths of an inch of pressure and image positioning.

Printing presses are configured as either sheet-fed presses or web presses. Sheet-fed presses print individual sheets of cut paper, whereas web presses print roles of paper. Presses are further configured with printing stations or units. Each unit prints a different color or provides other applications such as special coatings or varnish to enhance the look of the finished printed piece. Typical configurations are one-color, two-color, four-color, five-color, and six-color presses. However, printing presses can also be manufactured with more than six units for specialty printing purposes. Multiple units allow the press to print two or more colors on the substrate in one pass through the printing press. For example, a four-color press would have a separate unit for yellow, magenta, cyan, and black—these are the four colors needed for full-color printing. A six-color press may be used to add a fifth specialty color and then a varnish on the sixth unit. HiFi (high-fidelity) printing requires at least a six-color press.

Heidelberg Five-color Sheet-fed Press (Heidelberg)

These single or multiple unit configurations are available in sheet-fed and web presses as well as in all major printing processes. There is also a "combination" press. Combination presses are those with more than one printing process. For example, a press that uses both the flexographic and gravure printing processes is considered a combination press, as would be the case of a printing press that uses a combination of lithography and inkjet printing. Combination presses are sometimes used in package printing, printing requiring imprinting or addressing, and in security printing where documents must be produced that are difficult to copy or counterfeit. Scratch-off gaming tickets, such as lottery tickets, are produced on combination presses using a multiplicity of processes such as lithography, gravure, flexography, and inkjet printing, but are not limited to this combination of processes. These presses sometimes have as many as sixteen units because of the number of colors printed as well as the various layers of coatings needed to enhance security.

Every different printing process is characterized by its image-carrier or plate characteristics. They are also characterized by the formulation of their inks (or toners in the case of inkjet and electrostatic printing) and by the way each process is identified under magnification. In other words, if one skilled in the art of printing wanted to identify how a printed item was produced, this can be determined by viewing the printing under magnification of 12-power or more. Lithography has a different look than does letterpress, gravure has a different look than does flexography, inkjet printing has a different look than does electrostatic printing, and so on.

Goss Web Press
(Goss International)

Important attributes of a printed image are:
Register
Density
Dot gain
Trapping

Register is the relative positioning of the image on the substrate and the relative positioning of each ink layer over another. If the positioning of ink colors over each other is not accurate, the printed image will appear blurred.

Density is the intensity or visual strength of the ink that influences the color quality of the final printed image. In full-color printing, standards have been established for ink density when all other printing press variables are properly controlled. In other words, the density standard or target for yellow, magenta, cyan, and black are different in full-color printing. An instrument called a densitometer is used to measure the density of ink on a printed sheet.

Dot gain is the extent of growth that takes place in the size of a halftone dot or screen tint dot from the printing plate to the printed sheet. All printing processes—with the exception of electrostatic or electrophotographic printing—are subject to dot gain. It occurs because the ink is squeezed onto the substrate under pressure. Because the ink has a thickness to it, there is a tendency for the ink to spread under the pressure. In inkjet printing the ink is squirted onto the substrate and the force of the ink spot on the substrate causes the dot or spot to grow. Dot gain does not occur in electrostatic printing because dry toner particles are used that do not grow. Dot gain influences the look of a printed image. It is expected on the printing press and can be controlled in platemaking or in building digital files for printing. For example, if a dot gain of 30 percent is expected in the magenta ink being printed using the lithographic process, there are ways of reducing the size of the halftone dot or screen tint dot on the printing plate that will print the magenta ink.

Trapping is the extent to which one film of ink sticks to another when printing one ink-film over another. In full-color printing, the yellow, magenta, cyan, and black halftones must partially overprint each other to satisfactorily produce the resulting red, green, and blue colors needed in full-color printing. Ideally, 100 percent of an ink film will stick to the other. This often occurs when a wet ink film is applied to a dry one. However, in reality, because the inks are wet on a multicolor press, less than 100 percent of one ink film adheres to the other. If only 85 percent sticks or transfers, this is called 85 percent trapping. If 70 percent sticks, this is called 70 percent trapping. The degree to which trapping occurs influences the look of the final print. The percent of trapping too is measured with a densitometer.

Other components of a press that impact what the final print will look like include plate-to-blanket squeeze pressures on offset lithographic presses; ink film thickness on the press; the balance of fountain solution (water) and ink in lithographic printing; the pH and conductivity of the fountain solution; roller settings and roller hardness; the tension of the substrate going through the press in web printing; and more.

Traditional Printing Processes

Letterpress Printing - Letterpress printing is for the most part an obsolete process, but it was the mainstay of printing for several hundred years. Letterpress is printing from the surface of a raised image. It is also called relief printing, where the plate image is raised above the non-printing areas. It is the oldest of the present printing processes and has declined in use significantly over the past 50 years as more efficient, economical, and higher quality printing processes emerged such as lithography and flexography. Less than five percent of all printing in the United States involves the letterpress process and its use continues to decline. The typical letterpress plate is made of zinc, copper, a multi-metal formulation composed primarily of lead, or photopolymer. The demise of letterpress is directly related to the amount of set-up time required, the heaviness of the plates and the cumbersome nature of making plates, the cost of the plates, and image resolution limitations.

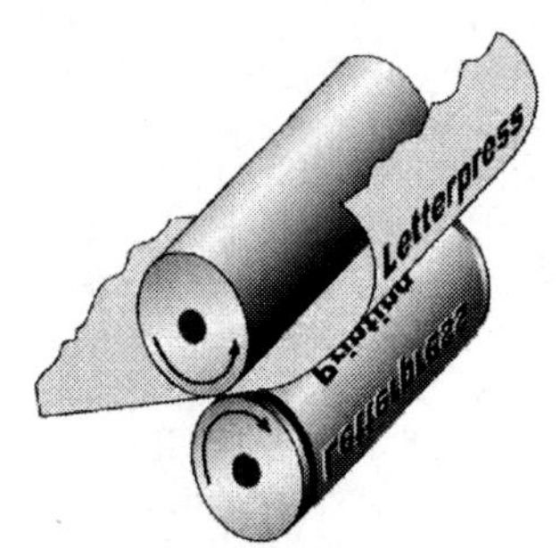

Letterpress Printing
(Tecstra Systems and International Paper)

In letterpress printing, ink is placed in the press ink fountain and then is distributed onto ink rollers. The ink rollers role the ink onto the raised image of the plate, and the plate transfers the image onto a substrate (usually paper). In this process, the image on the plate must be “wrong-reading” (or a mirror image) so that the printed image on the substrate will be “right-reading.”

Typical traditional products printed with the letterpress process include business cards, letterhead, proofs, business forms, posters, announcements, advertising flyers, brochures, imprinted items, and publications. However, with only few exceptions, today letterpress has been relegated, not to printing, but to the finishing processes of embossing, die cutting, and foil stamping.

Letterpress printing exerts variable amounts of pressure on the substrate dependent on the size and image elements in the printing. The amount of pressure per square inch—or “squeeze”—is greater on some highlight dots than it is on larger shadow dots. Expensive, time-consuming adjustments must be made throughout the press run to ensure that the impression pressure is accurate.

Letterpress printing uses type that is raised above (in relief) the non-printing areas. In traditional letterpress printing, letters were assembled into copy, explanatory cuts were placed nearby, line drawings were etched or engraved into plates—and all these were placed (composed) on a flat stone within a rigid frame called a chase, spaced appropriately with wooden blocks called furniture, and tightened or locked-up with toothed metal wedges called quoins.

There are three different types of letterpress printing presses: platen, flatbed, and rotary. The most common types of press used when the process was popular were the perfecting rotary press and the rotary letterpress typically used for newspaper and magazine printing.

Flatbed cylinder presses used either vertical or horizontal beds. The plate was locked to a bed that passed over an inking roller and then against the substrate. The substrate passed around an impression cylinder on its way from the feed stack to the delivery stack. Another way of describing this is that a single revolution of the cylinder moved over the bed while in a vertical position so that both the bed holding the substrate and cylinder moved up and down in a reciprocating motion. Ink was supplied to the plate cylinder by an inking roller and an ink fountain. The presses printed either one- or two-color impressions. Flatbed cylinder presses, which operated in a manner similar to the platen press, printed stock as large as 42 inches by 56 inches.

Flatbed cylinder presses operated very slowly, having a production rate of not more than 5,000 impressions per hour. As a result, much of the printing formerly done on this type of press was moved to rotary letterpress or lithography.

There were two types of rotary letterpresses: sheet-fed and web-fed. Web-fed rotary presses were at one time the most popular type used for letterpress printing.

Like all rotary presses, rotary letterpress required curved image carrying plates. The most popular types of plates used were stereotype, electrotype, and molded plastic or rubber. When printing on coated papers, rotary presses used heat-set inks and were equipped with dryers, usually the high-velocity hot air type.

Web-fed rotary letterpress presses were used primarily for printing newspapers. These presses were designed to print both sides of the web simultaneously. Typically, they printed up to four pages across the web; however, some of the newer presses printed up to six pages across a 90-inch web. Rotary letterpress was also used for long-run commercial, packaging, book, and magazine printing.

Lithography - Lithography is printing from a flat surface on which the image areas and non-image areas are on the same plane. The process is based on the principle that grease and water do not mix. The image and non-image areas are separated chemically in such a way that the image on the plate will accept greasy ink and the non-image areas will accept water and afterward reject ink. On a typical lithographic press there is an ink fountain and a water or dampening fountain. Ink is distributed from the ink fountain onto a set of ink rollers. Simultaneously, the water fountain distributes a dampening solution, primarily composed of water, to dampening rollers. The rollers dampen the plate before ink is applied to it. The water sticks to the non-image areas that were chemically treated to accept the water. The ink rollers then apply ink to the plate. Because the water on the non-image areas rejects the greasy ink, the ink will only stick to the image areas. The lithographic plate is typically made of aluminum, although other metals as well as paper and plastic can be used.

The inked images are then transferred to a rubber-like blanket that is wrapped around a cylinder that comes in contact with the plate cylinder. From the imaged blanket the image is transferred to the substrate being printed.

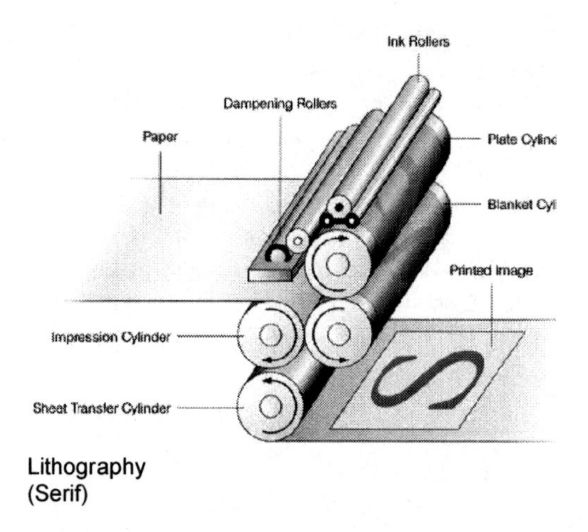

Lithography
(Serif)

The blanket performs three tasks. The first task is to allow a right-reading image on the plate to become right-reading on the substrate. If the blanket cylinder were not in place, the image would go from right-reading on the plate to wrong-reading on the substrate. With the blanket in place, the image goes from right-reading on the plate to wrong-reading on the blanket to right-reading on the substrate. The second function of the blanket is to reduce the amount of water that reaches the substrate. When printing on paper, moisture that absorbs into the paper causes problems such as paper distortion or dimensional instability. The third role of the blanket is to allow printing a large variety of substrates—regardless of texture in most cases. The blanket allows for a certain degree of compressibility. Therefore, when printing on rough-texture papers having peaks and valleys, the ink can be forced into the valleys of the paper.

Lithography is presently the most popular and widely used printing process for most printed products ranging from simple single-color printing to high-quality full-color printing. It is well suited for printing text and illustrations in short to medium length runs of up to 1,000,000 impressions. Approximately 50 percent of all printing in the United States is produced using the lithographic process and its use is slowly declining as digital processes improve in speed, size, and quality.

There are sheet-fed and web lithographic presses. On sheet-fed presses, the substrate is fed into the press one sheet at a time at a very high speed. Web-fed presses print on a continuous roll of substrate, or web, which is later cut to size.

There is also non-heat-set and heat-set offset lithographic printing. The difference between the two is primarily dependent on the type of ink used and how it is dried. Non-heat-set inks are typically dried via oxidation and absorption, such as in newspaper printing, and heat-set inks require special drying mechanisms on the press such as heating ovens, ultraviolet, infrared, or electron beam dryers. High-quality and glossy printing requires such drying devices.

Gravure - The gravure process has its origins in the early seventeenth century when the intaglio printing process was developed to replace woodcuts in illustrating the best books of the time. In early intaglio printing, illustrations were etched on metal, inked, and pressed on paper. Gravure, still also known as intaglio printing, makes use of the ability of ink to fill a slight depression on a polished metal plate. The basics of gravure printing are a fairly simple process that consists of a printing cylinder, a rubber-covered impression roll, an ink fountain, a doctor blade, and a means of drying the ink.

In principle, gravure printing can be thought of as the opposite of letterpress printing. Whereas letterpress prints from a raised image, gravure prints from a recessed image. In other words, in gravure printing the image area is beneath the plate surface and the non-image area is on the plate surface. The typical gravure plate is made of a large copper cylinder. Through chemical, electro-mechanical, or laser engraving processes, an image is etched or engraved into the copper cylinder in the form of microscopic wells or cells.

Ink in the gravure press ink fountain is applied directly to the copper cylinder and the ink will not only fill the wells but will also adhere to the surface of the cylinder. Ink is applied to both the image and non-image areas of the cylinder. A doctor blade made of hard rubber or plastic then passes over the cylinder and scrapes the ink off of the cylinder surface or non-image area. After this occurs, the substrate being printed comes in contact with the cylinder at high speed and under high pressure. As the paper is rapidly pulled off of the cylinder, a capillary action pulls the ink out of the cylinder ink wells, which represent the image area, and the ink is transferred onto the substrate. This all occurs at a high rate of speed.

Gravure printing involves high costs and a lot of time in preparing the plate cylinder. It is, therefore, economical for very long press runs where the cylinder does not have to be changed often. Typically, printing requiring tens of millions of impressions lends itself to the gravure process. Another unique aspect of gravure is its image quality potential. The process allows for smooth tone transitions from highlight tones to middle tones to shadow tones. Because of this, the process is also often used when extremely high-quality printing is required.

Gravure is a popular process for long-run publication printing, as well as package printing on non-paper or board substrates such as foils, plastics, cellophane, and other substrates that have little or no absorption. It is also a popular process for printing on specialty items such as wall coverings and linoleum and for producing synthetic wood grains on pressure-sensitive substrates. Gravure represents approximately 14 percent of all printing and is gradually declining in use.

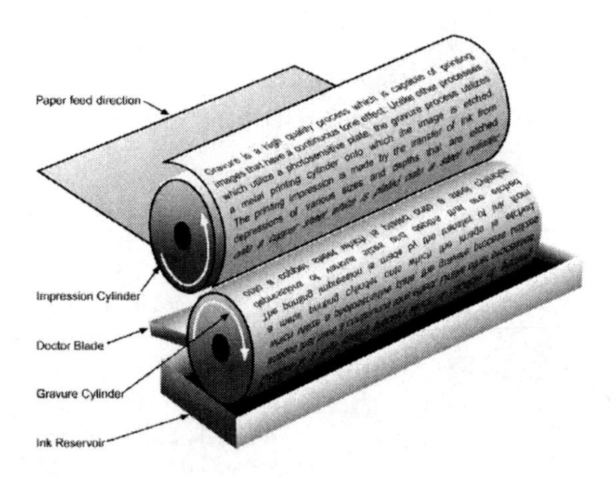

Gravure
(Tecstra Systems and
International Paper)

The dominant gravure printing process, referred to as rotogravure, employs web presses equipped with cylindrical plates as image carriers. A number of other types of gravure presses are currently in use. Rotary sheet-fed gravure presses, though rare, are used when high-quality pictorial impressions are required. They find limited use, primarily in Europe. Intaglio plate printing presses are used in certain specialty applications such as printing currency and fine art. Offset gravure presses are used for printing substrates with irregular surfaces or on films and plastics.

Today, almost all gravure printing is done using engraved copper cylinders protected from wear by the application of a thin electroplate of chromium. The cylinders used in rotogravure printing can be from three inches in diameter by two inch wide to three feet in diameter by 20 feet wide. Publication presses are from six to eight feet wide, while presses used for printing packaging rarely exceed five feet in width. Product gravure presses show great variation in size, ranging from presses with cylinders two inches wide that are designed to print wood grain edge trim to cylinders 20 feet wide that are designed to print paper towels.

Flexography - Like letterpress printing, flexography involves printing from a raised image on the plate. The difference, however, is that the flexographic plate is typically made of synthetic rubber or a photopolymer material. Some of the harder flexographic photopolymer plates print relatively sharp and produce high-resolution images. However, the softer synthetic rubber plates are not suitable for high-quality printing and are used for long-run imaging requiring one or two flat colors where image sharpness is not a critical concern. The technology of flexography has improved rapidly over the past decade, as has the quality. The process is popular for label printing; packaging; corrugated board printing; printing on non-paper substrates such as cellophane, plastic, polyester, foils, folding cartons, paper bags, plastic bags, milk and beverage cartons, disposable cups and containers, adhesive tapes, envelopes, food wrappers, and other substrates where there is little or no ink absorption. In recent years, flexography has also become popular for newspaper printing because the process lends itself to the use of water-based inks that do not rub off when handled. Flexography represents approximately 19 percent of all printing and its use is growing.

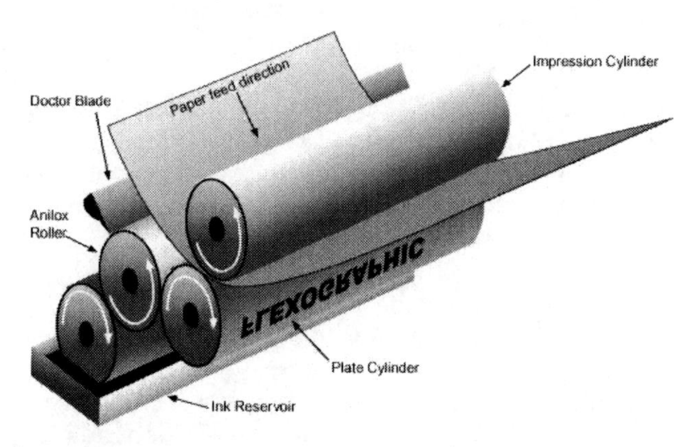

Flexography
(Tecstra Systems and
International Paper)

In the typical flexographic printing sequence, the substrate is fed into the press from a roll. The image is printed as the substrate is pulled through a series of stations or print units. Each print unit prints a single color. As with gravure and lithographic printing, various tones and shading are achieved by overlaying the four colors of ink—called process colors, these are magenta, cyan, yellow and black.

The major operations in flexographic printing are image preparation, platemaking, printing, and finishing. Image preparation begins with camera-ready (mechanical) art and copy or electronically produced art supplied by the customer. As in other processes, images are captured for printing by camera, scanner, or computer. Components of the image are manually assembled and positioned in a printing flat when a camera is used. This process is called image assembly. When art and copy is scanned or digitally captured, the image is assembled by a computer with special software. A proof is prepared to check for position and accuracy; when color is involved, a color proof is submitted to the customer for approval.

Flexographic plates are relief plates that come in contact with the substrate being printed. The plates are attached to a roller or cylinder for ink application. There are three primary methods of making flexographic plates: photomechanical, photochemical, and laser engraving.

The five types of printing presses used for flexographic printing are the stack type, central impression cylinder (CIC), in-line, newspaper unit, and dedicated four-, five-, or six-color unit commercial publication flexographic presses. All five types employ a plate cylinder, a metering cylinder known as an anilox roll that applies ink to the plate, and an ink pan. Some presses use a third roller as a fountain roller and, in some cases, a doctor blade for improved ink distribution.

Flexographic inks are similar to packaging gravure printing inks in that they are fast drying and have a low viscosity. The inks are formulated to lie on the surface of non-absorbent substrates and solidify when solvents are removed by drying devices. Solvents are removed with heat unless UV curable inks are used.

Screen Printing - Screen printing is probably the simplest of the major printing processes. The image to be printed is formed on a screen made of synthetic fibers over which a stencil is placed that represents the non-image areas. The area of the screen not covered by the stencil represents the image area because it is here where ink can pass through the screen onto the substrate. Stencils can be formed in a number of ways. One way is photographically by exposing negative or positive film to a photographic emulsion applied to the screen. When developed, the image and non-image areas are defined. Stencils can also be formed by applying pressure-sensitive stencil material on the screen or by "painting" a liquid stencil on the screen. Once the stencil is formed, the screen is brought in contact with the substrate, ink is placed on the screen, and a squeegee drags the ink over the stencil and the entire screen. The ink that is not blocked by the stencil will go through the screen and onto the substrate to form the printed image.

While screen printing is often a manual and slow process, there are automatic single and multi-color screen presses and even web screen presses. The screen-printing process lends itself to printing that does not require long runs because it is a relatively slow process. It is often used for printing T-shirts on nylon, cotton, and other textiles, as well as for printing short run-posters, bumper stickers, billboards, labels, decals, signage, electronic circuit boards, glass, leather, wood, ceramic surfaces, and paper. The advantage of screen printing over other printing processes is that the press can print on substrates of many shapes, thickness, and sizes. Screen printing represents approximately five percent of all printing and its use is growing gradually.

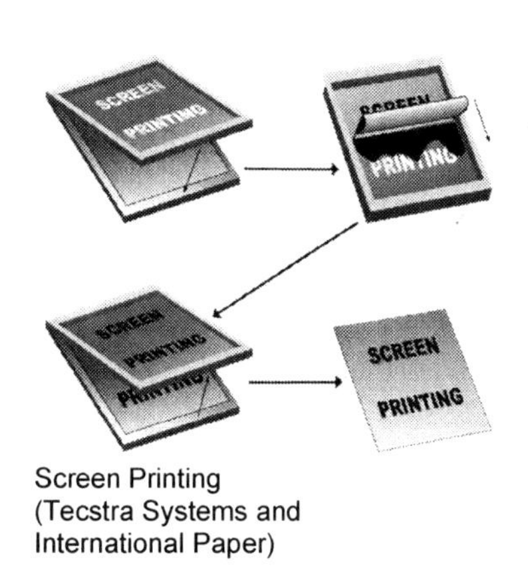

Screen Printing
(Tecstra Systems and International Paper)

A significant characteristic of screen printing is that a greater thickness of the ink can be applied to the substrate than is possible with other printing techniques. This allows for some effects that are not possible using other methods. Because of the simplicity of the process, a wider range of inks is available for use in screen printing. These inks are moderately viscous and exhibit different properties when compared to other printing inks such as offset, gravure, and flexographic inks, though they have similar basic compositions (pigments, solvent carrier, toners, and emulsifiers). There are five different types of screen printing inks including solvent based, water based, solvent plastisol, water plastisol, and UV curable.

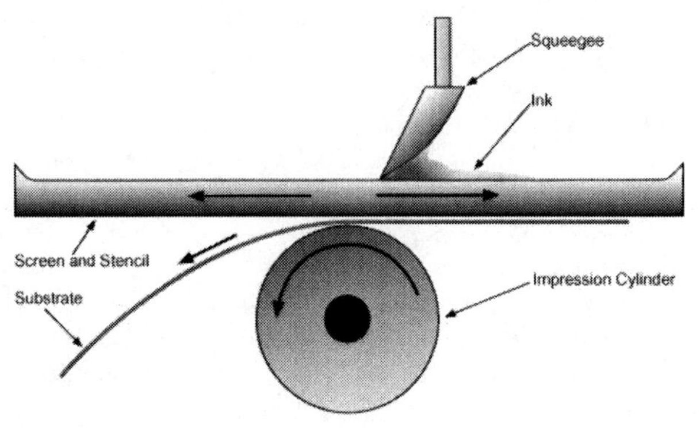

Screen Printing - Rotary
(Tecstra Systems and
International Paper)

The use of screen printing has grown slightly in recent years because production rates have improved. This has been a result of the development of automated and rotary screen printing presses, improved dryers, and UV curable ink. The major chemicals used include screen emulsions, inks, solvents, surfactants, caustics, and oxidizers used in screen reclamation. The inks used in this process vary dramatically in their formulations.

Screen printing consists of three elements: the screen which is the image carrier, the squeegee, and ink. The process uses a porous mesh stretched tightly over a frame made of wood or metal. Proper tension is essential for accurate color registration. The mesh is made of porous fabric or stainless steel. The diameter of the threads and the thread count of the mesh will determine how much ink is deposited onto the substrates. The stencil, produced on the screen either manually or photochemically, defines the image to be printed. This is analogous to the image plate in other printing processes.

Many conditions such as composition, size and form, angle, pressure, and speed of the blade (squeegee) determine the quality of the impression made. At one time most blades were made from rubber that is prone to wear, subject to edge nicks, and has a tendency to warp and distort. While blades continue to be made from rubbers such as neoprene, most are now made from polyurethane that can produce as many as 25,000 impressions without significant degradation of the image.

If an item is printed on a manual or automatic screen press, the printed product will be placed on a conveyor belt that carries the item into a drying oven or through a UV curing system. Rotary screen presses feed the material through the drying or curing system automatically. Air-drying of certain inks, though rare, is still sometimes used. The rate of screen-printing production was once dictated by the drying rate of screen-printing inks. Due to improvements and innovations, the production rate has greatly increased. Some specific innovations that affected the production rate and have also increased screen press popularity include the following: development of automatic presses versus hand-operated presses, which have comparatively slow production times; improved drying systems, which significantly enhance production rate; development and improvement of UV curable ink technologies; and development of the rotary screen press, which allows continuous operation.

There are three types of screen-printing presses: the flatbed (which is the most common), cylinder, and rotary. Until relatively recently, all screen-printing presses were manually operated. Now, however, most commercial and industrial screen printing is done on single and multicolor automated presses.

Combination Printing

Combination printing involves printing presses that use two or more printing processes. For example, security printing—such as scratch-off lottery tickets—typically employs printing presses having three or four processes such as lithography, gravure, flexography, and inkjet printing. Package printing often involves combination presses to create different effects of solids, screens, or foil finishes.

Digital Printing

Digital printing has come of age in the early part of the twenty-first century. Digital presses now produce images that rival offset lithography and lower cost-per-image pricing makes them a viable printing alternative.

Digital printing presses are presses that are able to accept jobs directly from the computer and produce a printing plate or image a cylinder. They allow taking on jobs and run lengths that were previously just not cost effective.

Digital printing represents a growth area for many graphic communication companies. Digital printers do mostly digital printing, but some do traditional printing as well. They are also producing a growing amount of full-color printing to complement the large volume of black-and-white printing traditionally associated with digital printing through companies such as service bureaus and corporate data centers where large volumes of bills, statements, policies, and reports are printed.

In these market segments, while black-and-white printing is predominant, there is a move to incorporate color in their documents. In some cases, companies print variable black text on a shell that has been pre-printed in color using offset presses. Direct mail printers who previously printed almost exclusively in black when it came to digital printing have now incorporated full-color digital printing into their production mix.

Most of the digital printers are small. The majority has less than five employees, and only a few hundred of them have more than ten employees. In an industry where many firms are still fighting to survive, digital printers are thriving.

Digital printers produce a variety of printed products. Although they all do some color pages, black-and-white printing still accounts for the majority of pages for most companies in this segment. Books and manuals make up the largest page volume, and these products are printed predominantly in black and white and sold at a relatively low cost per page. The revenue makers and growth areas are color applications, such as brochures, other sales collateral, and direct mail printing. Individual companies have also developed specialized color markets, such as business cards, short-run posters, and calendars.

Digital printers serve customers from a variety of industries with some of the most popular being business services, financial services, retailers, non-profit organizations, and education and government organizations.

Firms classified as digital printers are more likely than firms in other segments to have color digital printing equipment in the speed range from 60-ppm upwards. In proportion to their numbers, they are far more likely than others to have black-and-white sheet-fed presses at 106-ppm or higher. Additionally, digital printers, more than any other graphic communication segment, have plans for expanding their capabilities with top-of-the-line color digital presses and high-speed monochromatic systems.

A key consideration that has allowed the digital printers to grow in an environment where other printing companies cannot is the availability of affordable digital color. The printed products that these companies produce (business-card printing, postcards, brochures, posters, and direct-mail pieces, among others) have been impractical to print in color in short runs using traditional printing equipment. But with color digital printing costs dropping below ten cents per letter-size page, the economics have become attractive. Many color devices now have operating costs that are well under ten cents per page, and a few are under five cents per page. There is the expectation that page costs will continue to drop. For the immediate future, however, the cost per color page for digital printing is still too high to compete with offset printing for long runs. But for runs of up to between five thousand and ten thousand pages, digital printing has become increasingly cost-effective. As the cost of digital printing continues to drop, the range of applications that can be economically printed using digital technology will expand.

Most digital printers are not yet primarily focused on variable data printing (VDP), although almost all of them own software to handle variable data. Only a few are variable data specialists, but the number is growing, as here is where digital printing has capabilities that traditional printing does not.

For the last decade, vendors of digital printing equipment have been promoting VDP because, in certain applications, variability could justify the high cost of digital color pages. However, digital printing was not able to compete with offset lithography for the printing of static pages, except in extremely short runs. With the emergence of digital printers and the advantages that such printing provides, the era when digital printing was confined to exceptionally high-value niches has drawn to a close. There will continue to be compelling reasons for doing VDP, but there will be plenty of other work printed on digital devices as well. Yet VDP will continue to grow, along with short-run color. After years of much promise but minimal payback, VDP will eventually achieve widespread use.

Two or three trillion pages are printed in North America each year, mostly on offset equipment, and much of that volume will eventually be a candidate for digital printing. For at least half of those pages, printing cost is the primary obstacle to shifting from offset to digital. And the cost of digital printing is finally reaching the level at which it can compete for a fraction of those trillions of pages. Replacing offset printing with digital printing will be the real growth area for digital printing. However, success in digital printing requires a good marketing strategy, an efficient workflow, and good customer service.

Digital printing has captured the attention of the printing industry and is being increasingly demanded by the industry's customers including advertising agencies, design studios, publishers, and marketing firms. Some of the broad advantages of digital printing include economical short-run color, on-demand printing, personalized (variable data) printing, reduced or no inventory of printed products, relatively inexpensive press technology, digital workflows, no printing plates in some cases, and a clean operating environment.

There are a growing number of digital printing presses impacting the present mode of print production and distribution with the promise of more being developed in the future. Such systems appeal to certain niche markets such as on-demand, variable data, short-run, black and white, and color printing of commercially acceptable to high-quality printing. However, the promise of such technology is in commercial printing, presently dominated by traditional processes such as lithography. This promise will be realized as digital printing systems continue to improve in the quality of the printed image, the size of sheet or web accommodated, and the speed of production. Digital printing currently represents approximately seven percent of all printing and is growing.

Many of the digital printing processes are also "pressureless" processes. For example, electrostatic/electrophotographic and inkjet printing do not use traditional wet printing ink but use toner (dry toner in the case of electrostatic/electrophotographic printing and wet toner in the case of inkjet printing) that is directed to the image area on the substrate via electrostatic charges. Hence, there is no printing plate squeezing ink onto a blanket or directly onto the substrate.

On-demand printing is the same as JIT (just in time) printing where printing is provided at the same time (or close to the same time) that it is requested. Electronic systems are used for this. Variable data printing involves a print run in which each page or sheet is printed with different information. The difference can be in what is included in words and numbers, and also in pictures or other graphics. Variable data printing is often done to "personalize" what would otherwise be non-personal direct mail advertising. Scratch-off lottery tickets are another form of variable data printing where each ticket has different numbers and identification codes but is printed as part of a single press run.

Digital Printing Engines

There are different printing engines that drive digital presses. Digital systems consist of inkjet, electronic, electrophotographic, magnetographic, ion deposition, light emitting diode (LED), liquid crystal shutter (LCS), electron beam imaging (EBI), thermal, and electrostatic printing. These are all processes—used mainly for short runs and printing variable or personalized information—in which data representing the images are in digital form.

Inkjet Printing - Inkjet printing is used mainly for variable printing such as addressing, coding, computer letters, sweepstakes forms, and other personalized direct mail advertising. During the inkjet printing process, microscopic droplets of ink are squirted onto a substrate from a print head containing one or more nozzles. The two main types of inkjet printers are continuous drop and drop-on-demand. In continuous drop inkjet printing, a continuous stream of ink droplets is generated through a nozzle under constant pressure. In the drop-on-demand process, ink drops are expelled from the nozzle only when needed. There are three types of drop-on-demand inkjet printing: piezoelectric, where a piezoelectric crystal produces an electric charge that causes the ink drop to be expelled; bubble jet/thermal liquid ink, where an electric charge is applied to a small resister causing a minute quantity of ink to boil and form a bubble that expands and forces the ink droplet out of the nozzle; and solid ink, which involves a solid wax-based ink that melts quickly and solidifies on contact with a substrate.

In continuous inkjet and drop-on-demand processes, the ink droplets are deflected to the image area via electrostatic charges. Unneeded droplets are not charged and are deflected into a gutter for recycling. A disadvantage of inkjet printing is that the inks are water-soluble and can easily smudge when subjected to moisture (applying a protective coating can prevent smudging).

Magnetographics - Magnetographics is a plate or cylinder "pressureless" printing process similar in principle to electronic printing. The difference is that the photoconductor and toners are magnetic, and a device similar to a printing press is used. The process uses an array of thousands of heads to create a magnetic image on a drum. As the drum rotates, the magnetic image picks up toner. Excess toner is removed magnetically or by vacuum as the image is transferred to a substrate through a transfer roller. The image is fused to paper by heat that melts the toner, after which the magnetic image is removed from the drum by an erasing head. Through this sequence, the printing cylinder is charged, imaged, and toned for the first impression and recharged and toned for succeeding impressions. Magnetography is a short-run process with a break-even point with lithography of about 1,500 impressions. Its main advantage is ease of imaging with digital data. Its limitations are slow speed, absence of light-colored transparent toners (so it is not yet suitable for process color printing), and high toner costs.

Ion Deposition Printing - Ion deposition printing is similar to other electronic printing systems. The process, having few moving parts, produces dots at high speed using high-density ion currents. Images are produced directly from electrical pulses and involve a substrate passing—under pressure—between the imaged drum and a transfer roller. It consists of four simple steps: (1) an electrostatic image is generated by directing an array of charged particles (ions) from a patented ion cartridge toward a rotating drum consisting of very hard anodized aluminum maintained at a constant temperature; (2) a single component magnetic toner is attracted to the image on the drum as it rotates; (3) the toned image is transfixed to plain paper; and (4) after scraping the very small amount of toner (0.3 percent) left on the drum, it is ready for re-imaging. Ion deposition printing is used for printing invoices, manuals, letters, and proposals, as well as for specialty printing of tags, tickets, and checks. A disadvantage to this process is that images are easily rubbed. A new system is being developed using new materials capable of producing high quality continuous-tone four-color process images.

Light Emitting Diode (LED) Printing - Light Emitting Diode (LED) printing uses a light source to create an image on a photoconductor. The system employs an array of several thousand light-emitting diodes that provide direct imaging with no moving light beam. In this process, light-emitting diodes are switched on and off to create an image on a charged drum.

Liquid Crystal Shutter (LCS) Printing - The Liquid Crystal Shutter (LCS) process is similar to the Light Emitting Diode process; the difference is that LCS writes an image onto a photoconductor through staggered liquid crystal "shutters" in place of diodes. The system uses a high-intensity fluorescent bulb for light exposure during which electrical signals open and close shutters depending on the areas to be imaged.

Electron Beam Imaging (EBI) Printing - Electron Beam Imaging (EBI) differs from ion deposition printing in that imaging is performed by controlling a beam of electrons instead of ions. The system uses high voltage and high frequency to excite free electrons onto a rotating drum. The electrons attract toner particles in the same way that laser printers do. High-pressure rollers are used to fuse the toner to the paper.

Thermal Printing - Thermal printing is widely used for non-impact color printing. There are two thermal printing processes: direct thermal and wax thermal transfer. In direct thermal printing a specially coated thermographic paper is exposed to a head that reacts by turning the exposed areas into different colors, either black or blue. In the wax thermal transfer process the same chemical principle is used. However, a special coated ribbon releases pigmented material. Characters are printed one at a time by a print head that moves back and forth.

Electrostatic Printing - Electrostatic printing is similar to "xerographic" or electrophotographic printing. The process uses toner particles to form an image, not from digital data but from an analog original. Unlike electrophotographic systems, there is no print drum. Toner particles are attracted directly to the paper through controlled conductivity. No optical system is used. The entire copier glass is exposed at once and an electrostatic charge is directly deposited onto the paper. Electrostatic printing systems use liquid or dry toners. In systems that require liquid toners, the toner is fused to the paper through hot air. When dry toners are used, the toner is fused to the paper via pressure.

Electrophotographic Printing - Electrophotographic printing is similar to high-speed copier systems operating on the principles of “xerography.” It is the process used in many laser printers and copiers that operate from digital data. It uses an electrostatic photoconductor that is charged by a corona discharge, imaged by a moving laser light beam modulated by digital signals from a PostScript-based digital imaging system. In this process the laser beam is focused on a rotating mirror that deflects the beam through a focusing lens that forms a latent image on a photoconductor. Present systems work well for single or spot color specialty printing.

Commonalties Among Digital Printing Systems
Today's systems, while being different in many ways, also have many commonalties. For example, most are on-demand printing systems using digital technology; they all address the 50 to 20,000 niche market; and they all provide perfecting or duplexing capabilities and require short or no make-readies. Many require no warehousing of finished products because all products are produced one at a time and shipped to customers upon completion. Those systems using dry toners now approach offset lithographic quality, as do those using wet toners. Most of the systems provide variable imaging capabilities.

Wide Format Digital Printers
Wide format digital printers provide individuals and companies with access to inexpensive large-sized prints. There are a growing number of companies manufacturing systems that produce full-color digital prints ranging in size from 36 to 54 inches. However, systems that produce larger sizes are also available. There are two components to these systems: a large format printer and a RIP.

Large format digital printing systems print on a variety of substrates including paper and mylar and use engines represented by inkjet, electrostatic, or thermal wax transfer technology. This market is growing rapidly and finding its niche in the "quick printing" and on-demand printing industry segments. Users of this technology also include advertising agencies, screen printers, service bureaus, and in-plant printing departments of many businesses. The applications are numerous and include items such as art-on-demand, backlit signage, transportation advertising displays, engineering drawings, maps, murals, posters, window graphics, and more.

Wide Format Digital Printer
(Kodak Polychrome Graphics)

Digital Printing Presses
With the introduction of the first models in the early-to-mid 1990s, digital color printing presses have had a major impact on the printing industry sooner than expected. It typically takes ten years, and sometimes much longer, from conception to implementation and wide acceptance of an idea. It took desktop computers less than five years to become common technology in the graphic arts.

The digital evolution of color printing has accelerated in recent years beyond the imagination of color theorists and technicians. The paradigm that printing technology depreciates over a ten-year period has been replaced with a scientific revolution reducing the cycle of technological transition in the graphic arts from a decade or more to a few years or even months.

The technology of computer-to-film, bypassing the laborious tasks of graphic arts photography and image assembly, was a monumental step toward automating the printing processes in the late 1980s and early 1990s. Computer-to-plate technology, and its popularity in the 1990s, was an extension of this technology. However, computer-to-press technology provides the promise of completely integrated printing systems putting the author, copywriter, and artist in the position of producing finished printed products.

The Xerox DocuTech represented the first wave of such technology in the production of black-and-white printing. However, soon after DocuTech's introduction in the late 1980s, other manufacturers saw the future of direct-to-press technology in color markets and nearly all systems afterwards addressed this demand. Heidelberg, the first to introduce a direct-to-press color system in its GTO-DI, was quickly followed by Indigo, Xeikon, Agfa, and others. By the year 2000, Heidelberg had already introduced several generations of direct-to-press systems. Today several companies are producing digital color printing systems to compete with commercial printing presses.

Much of the accelerated development in this area is the result of mergers, joint ventures, and acquisitions where companies acquire existing technology rather than reinventing it. Thus, two or more technologies are brought together to create improved systems, speeds, and technological advances.

Digital printing has resulted in a wide range of new companies serving the industry's equipment needs. While traditional companies such as Heidelberg and MAN Roland have entered the digital arena, companies including Xerox, Canon, Hewlett Packard, IBM, Océ, Ricoh, Xeikon, Scitex, and others all are now suppliers of printing technology. These companies provide not only hardware but also intangibles such as software and digital front-end workflow systems. Digital workflow strategy and production workflow processes are as important as hardware in driving printer production and productivity.

On the hardware side, systems that handle a greater variety of substrates, including very lightweight paper, have been developed. The concept of "universal copier/printer" devices has been developed. These are devices that output color and monochrome pages at competitive costs with dedicated color and monochrome printers.

Adding value by integrating all services from front-end to printing to finishing is key to digital printing, as is flexibility and process improvement. A modern printing operation today also uses the World Wide Web as a common business tool for receiving information and for executing related business functions. Innovations in RIP, server, and workflow technology are making it easier to integrate digital devices into almost any printer's array of output services.

Digital printing has caught up with offset lithography in quality of output, and the digitally driven convergence of the two output technologies is already underway. Within the range of applications that they were designed for, digital presses compare with offset in color in such a way that even trained eyes can no longer distinguish the differences. Digital output systems now complement conventional offset lithographic equipment. They do not displace offset, as digital presses cannot yet deposit ink or coatings, print complex items such as magazines and die-cut packaging, or turn litho-sized papers into products at many thousands of sheets per hour. However, this is sure to change in the near future.

Digital presses, for the most part, use toners (dry and wet) in place of traditional printing inks. The exceptions are those lithographic presses that use digital technology for imaging plates and then print in the traditional offset manner.

Digital Copiers, Presses, and Printers

Digital Color Printers and Copiers

Color copiers represent the broadest range of manufacturers and features. Once relegated to the office environment, new digital color printers have become ubiquitous in printing companies as well. They employ basic laser imaging technology to charge an image on either drums or belts, from which the developed CMYK image is ultimately transferred to paper. Any project that demands process, full-bleed color in runs of up to 5,000 can be imaged on these high-resolution, toner-based devices.

Digital Color Production Presses

The NexPress 2100 and Xerox DocuColor iGen3 represent a breakthrough for commercial printers who want to bring short run/JIT, fast turnaround, web enablement, and personalized printing into the press department. Although there are individual differences, these toner-based presses are fast (up to 6,000

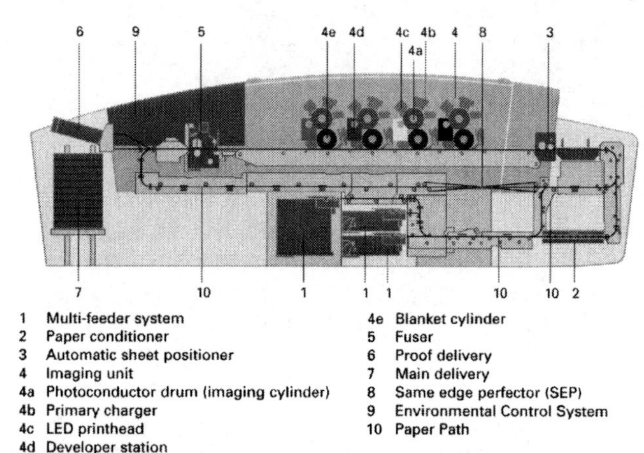

| Fig. 5.20 | Schematic diagram of the NexPress 2100

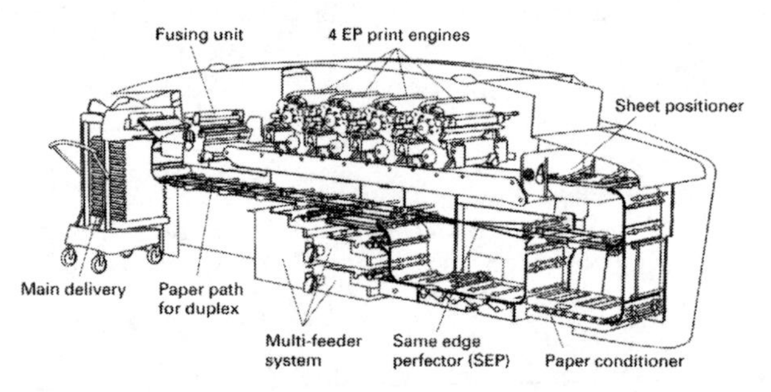

| Fig. 5.21 | Schematics of the NexPress 2100

NexPress
(NexPress and Océ "Digital Printing")

impressions per hour) and feature extremely high print quality (600 dpi). They run smooth and textured papers in a range of sheet sizes and basis weights—from 16 lb. bond to 110 cover—with the ability to mix stocks in a single run.

Xeikon-engined Printers

A toner-based web press, the Xeikon-engined printer feeds its web through a series of drums, each charged with the image and each applying one process

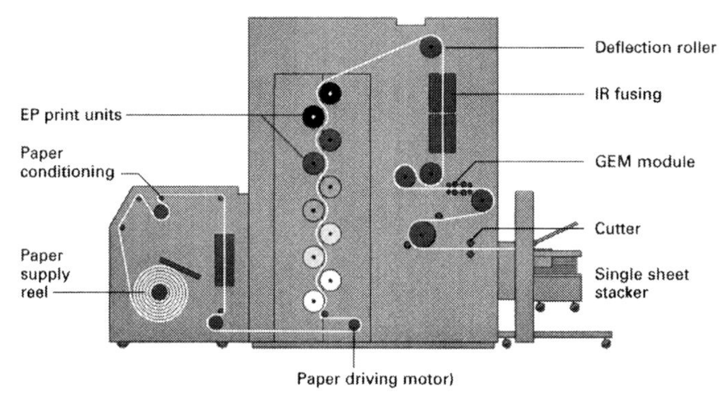

Xeikon Printing Units
(Xeikon and Océ "Digital Printing")

color. Process color toner is fused to the sheet with adjustable heat and pressure; changing the heat and pressure levels results in more or less gloss in the toner. An on-line cutter trims pages to length. For optimum performance paper must be scripted in order to establish set points on the equipment for heat and pressure as well as other key characteristics. This high-resolution duplexing printer has full variable data capability—meaning that some or all of the text or images can be changed from one document to the next.

Xeikon 5000
(Xeikon)

HP Indigo Presses

These presses use one imaging drum and patented liquid ElectroInk. Both drum and ink are charged; ElectroInk adheres to the image area on the drum and a blanket transfers the image to paper. No ink is left on the blanket. The plate charge is cleared and the process is repeated for subsequent colors. This process supports fully variable data and very high-resolution images. The HP Indigo

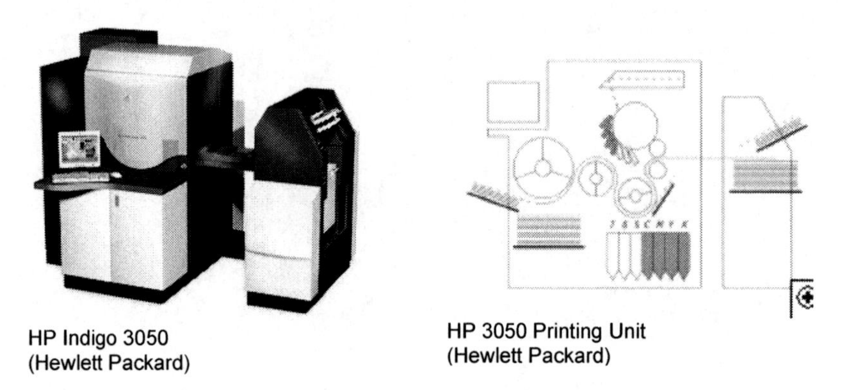

HP Indigo 3050
(Hewlett Packard)

HP 3050 Printing Unit
(Hewlett Packard)

press can print on a wide range of substrates, but for optimum performance, most papers must be pre-treated to improve toner adhesion.

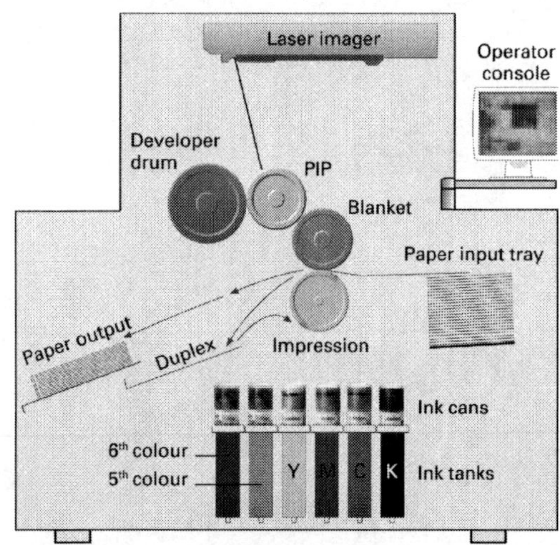

HP Indigo TurboStream Printing Unit
(Xeikon and Océ "Digital Printing")

DI Presses

These sheetfed presses work like offset presses with an electronic twist: directed by digital data, pre-mounted plates are imaged with a laser right on the press, reducing make-ready time to minutes. Ideal for process color jobs from 500 to 10,000, this equipment produces high-resolution offset images using traditional inks and is ideal for promotional and sales literature, publications, and even packaging.

Heidelberg Quickmaster DI
(Heidelberg and Océ "Digital Printing")

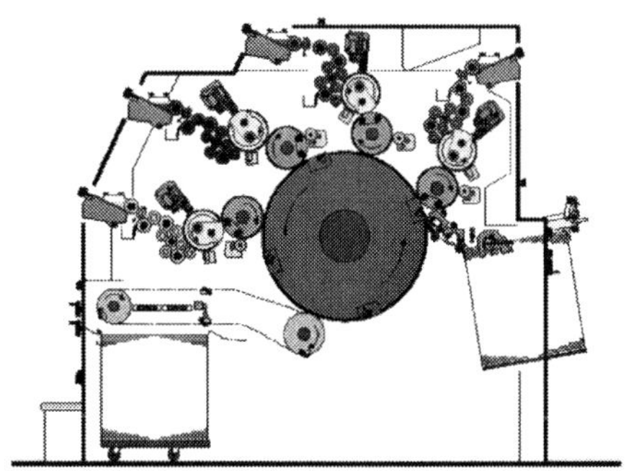

Heidelberg Quickmaster DI Printing Unit
(Heidelberg and Océ "Digital Printing")

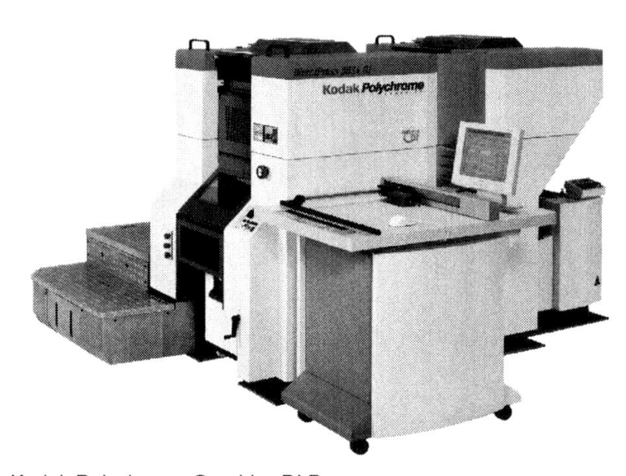

Kodak Polychrome Graphics DI Press
(Kodak Polychrome Graphics)

High-Speed and High-Volume Digital Press

High-speed and high-volume digital presses are used for documents requiring hundreds of thousands or even millions of copies. They have variable data capabilities and are very popular for transactional documents such as telephone bill, cable television bills, utility bills, and much more. They often have the capabilities of producing full-color images along with personalized messages directed specifically to the recipient. Océ Printing Systems is a major producer of such high-speed and high-volume digital presses. However, there are other manufacturers of such devices as well.

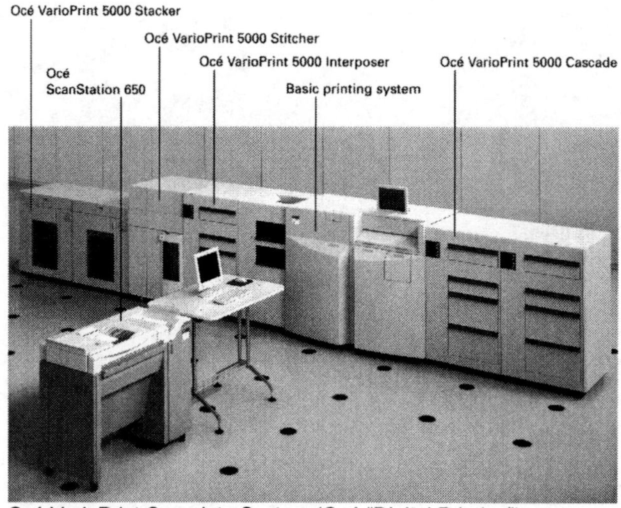

Océ VarioPrint Complete System (Océ "Digital Printing")

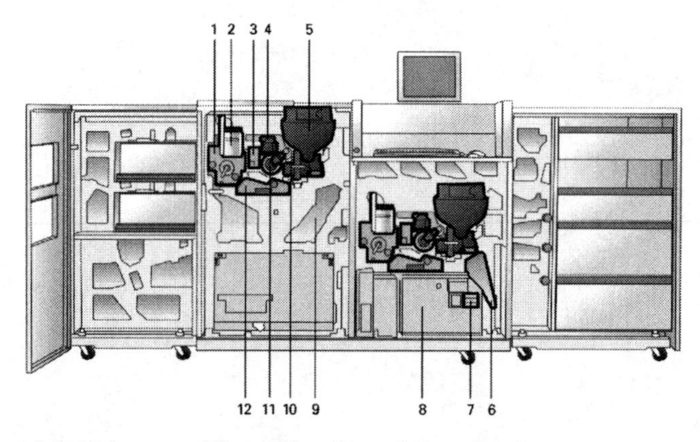

Océ VarioPrint Printing Unit (Océ "Digital Printing")

Océ VarioStream
(Océ "Digital Printing")

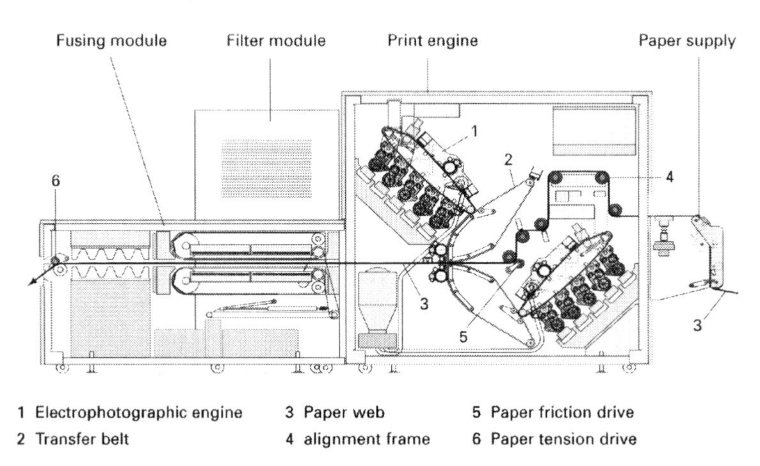

Océ VarioStream Printing Unit
(Océ "Digital Printing")

Inkjet Printers

Common in homes and office, inkjet printers today are capable of brilliant color reproduction with amazingly high resolution. Each requires a paper surface with sufficient ink holdout to reproduce a sharp, saturated image.

Summary of Leading Digital Press Characteristics

Digital Color Printers and Copiers

- Laser imaging technology to charge an image on drums or belts
- Developed CMYK image is transferred to paper

• Full-bleed color in runs of up to 5,000
• High resolution
• Toner-based devices

Digital Color Production Presses
• Short run/JIT
• Fast turnaround
• Web enabled
• Personalized printing
• Toner-based presses (up to 6,000 impressions per hour)
• High print quality (600 dpi)
• Smooth and textured papers in various sizes and weights from 16 lb. bond to 110 cover
• Can mix stocks in a single run
• Delivers collated sheets at the end of the press

Xeikon-engined Printers
• Toner-based web press
• Feeds web through a series of drums
• Each drum charged with the image
• Each drum applies one process color
• Process color toner is fused to the sheet with adjustable heat and pressure; changing the eat and pressure levels results in varying gloss in toner
• On-line cutter trims pages to length
• For optimum performance paper is scripted to set points for heat and pressure
• High resolution
• Duplexing
• Variable data capability

HP Indigo Presses
• One imaging drum
• Patented liquid ElectroInk
• Drum and ink are charged
• Ink adheres to the image area on the drum
• A blanket transfers the image to paper
• No ink is left on the blanket
• Plate charge is cleared and the process is repeated for each colors
• Variable data
• High-resolution images
• Prints on a wide range of substrates
• For optimum performance, paper is pre-treated to improve toner adhesion

DI Presses
• Sheet-fed presses
• Work like offset press with an electronic twist

- Directed by digital data
- Pre-mounted plates are imaged with a laser on the press
- Reduces make-ready time to minutes
- Ideal for process color jobs from 500 to 5,000
- Produces high-resolution offset images using traditional inks

Inkjet Printers

- Common in homes and offices
- Capable of brilliant color reproduction
- High resolution
- Requires paper surface with sufficient ink holdout for a sharp saturated image

Some Specific Digital Press Manufacturers and Press Features in the Early 21st Century

Here we list some of the systems available but by no means all of them. These descriptions are meant to cover some of the common features of digital printing presses.

Nipson

The Nipson 8000 is a black-and-white press that offers the ability to print on a variety of materials and makes use of the latest developments in print engine components and toner technology to provide low maintenance and consumable costs. It accepts media widths up to 20.5 inches with weights up to 100 pounds. It is capable of printing up to 229.5 feet per minute, with resolutions up to 600 dpi.

ADAST

The DI (direct imaging) family of ADAST printers, which includes the 547 and the 557, are 15 by 20.5 inches and use computer-to-plate technology produced by Presstek. Both presses feature offset-quality printing and have competitive price points for runs ranging from 500 to 5,000 prints. They use a unique design that declutches the plate cylinder to allow imaging while the rest of the press is idle. This reduces press wear and allows for high-speed imaging. They use no chemicals for plate development, making them environmentally safe. The 547 uses four colors, while the 557 uses five colors.

IBM Infoprint Line

The IBM Infoprint 3000 is a continuous form, production printer that can print up to 344 duplex impressions per minute for a monthly maximum duty cycle of eight million impressions. It can accept 240 or 300 dpi data streams and print them at 480 or 600 dpi.

The Infoprint 4000 family of printers offers a broad range of capabilities with expanded functionality and user-friendly control for highlight color, MICR, and direct mail applications. The Infoprint 4100 offers continuous form printing with 19-inch wide print line for digital publishing and statements printing. Speeds are up to 762 two-up, duplex letter or 718 two-up, duplex A4 impressions per minute with 480x600 dpi resolution. Simplex and duplex models are available.

The Infoprint Color 130 includes a RIP server, a high-performance press server, and a full-color digital printer designed for variable data printing. It prints up to 138 full-color letter-size impressions per minute with a monthly maximum duty cycle of 700,000 impressions. Designed for variable data, full-color printing, the Infoprint Color 130 Plus allows each copy of the print run to feature individualized images and data. It has classic screening at 600 dpi with up to 300 lpi screen resolution on fully variable pages.

The Infoprint 70 Plus is a cut-sheet printer that prints up to 70 impressions per minute and up to 600,000 impressions per month. It features print quality up to 600 dpi and offers flexible finishing capabilities. Offering print speeds of up to 85 impressions per minute, 600x600 dpi printing resolution, and full-function duplicating capabilities, the Infoprint 2085 is a high-speed cut-sheet printing solution. The Infoprint 2105 offers print speeds of up to 105 impressions per minute, 600x600 dpi resolution, and full-function duplicating capabilities. The Infoprint 2000 is a customizable production solution that supports 240, 300, and 600 dpi and automatic conversion and has up to 8,000 sheets on board. It can print up to 110 impressions per minute.

Canon

Canon's digital printer line features several models, including the CLC 5000+, the CLC 3900+, the imagePROGRAF W7200 and W7250, and the imagePROGRAF W2200.

The Canon CLC 5000+ has a 4000-sheet LTR-size paper deck or can come with a 2,000-sheet oversize paper deck. The CLC 5000+ is a new version of the CLC 5000 high-volume, 50 page-per-minute production color laser copier/printer. The CLC 5000+ has the same architecture as the CLC 5000, with the added ability to automatically duplex up to 60-pound cover to 90-pound index media. It uses four-drum technology and the Color Automatic Image Refinement System for resolutions up to 800x400 dpi.

The CLC 3900+ has a 4,000-sheet LTR paper deck or can come with a 2,000-sheet oversize paper deck. It is a new version of the CLC 3900 39 ppm, entry-level production color laser copier/printer. The CLC 3900+ features the ability to automatically duplex up to 60-pound cover to 90-pound index. The CLC 3900+ features all of the technologies built into the CLC 5000+, but at a lower price.

HP Indigo
The HP Indigo Press 1000 features high-resolution imaging, up to six-color printing—including spot and fluorescent colors—and virtually unlimited substrate selection. Printing speeds reach 2,000 full color 8.5x11-inch A4 images per hour or 8,000 single-color 8.5x11-inch A4 images per hour. The unit uses a liquid ink technology to produce offset-quality printing at 175 lines per inch.

The HP Indigo Press 3000 series is specifically designed for production environments and delivers up to seven-color printing and high-definition images. Throughput reaches 4,000 four-color 8.5x11-inch A4 images per hour or 16,000 single-color 8.5x11-inch A4 images per hour.

The HP Indigo Press w3200, designed for high-volume commercial printing, direct mail, and publishing applications, is a seven-color, web-fed unit that features a print rate of 8,000 8.5x11-inch A4 color images per hour. The liquid ink system delivers 800x800 dpi resolution, at 180 lines per inch.

MAN Roland
MAN Roland's DICOweb is an offset system that prints without plates. It uses an imaging cylinder that is laser imaged and then erased so the press can print directly from digital data without the need for plate changes. It can print 12x20.5-inch web width with user selectable cutoffs and can produce up to 20,000 8-page copies per hour.

KBA
Koenig & Bauer AG's (KBA) 74 Karat has a maximum resolution of 2,540 dpi, with a print size ranging from 11.75x8.25 inches to 29x20.5 inches. It has a 15-minute make-ready time from run to run and can produce up to 10,000 sheets per hour.

NexPress
The NexPress 2100 digital production color press, a joint venture between Heidelberg and Kodak, combines the reliability, durability, and high image quality of an offset press with the flexibility of a digital printer. It is a sheet-fed, full-color, auto-perfecting system that produces 2100 full-color, A3+ sheets per hour at 600 multi-bit dpi.

Heidelberg Digital Trio
The Heidelberg Digimaster 9110 Network Imaging System offers rapid image processing and a 110-page per minute print engine. It is a unit that can be used for duplication, network publication printing, and printing transactional documents. It has up to six paper supplies, a maximum page size of 11x17 inches, and finishing options that include booklet making.

The Digimaster 9150i, which enables 150-image-per-minute digital black-and-white printing, is an extension of the Digimaster 9110 print engine. It provides more flexibility to configure the press to meet specific needs and is faster.

Heidelberg has also positioned the Speedmaster 74 DI as a solution in mid-size 28-inch format. Features include 200 line screen, five-color printing, 19.5x28-inch image area, aqueous coating, and the ability to accept 70-lb. text to 18-pt. board. It is designed for runs of 500 to 20,000.

The Heidelberg Quickmaster DI 46-4 Pro has several enhancements that include a new control console and operating interface based on the company's CP2000 concept, a stream-feeder that allows the imprinting of a fifth or sixth color, and an optional infrared dryer.

Xerox

The DocuColor iGen3 Digital Production Press delivers short run, four-color printing at 100 pages per minute. The DocuColor 6060 Digital Color Press prints 60 full-color pages per minute and can print on a wide array of paper and substrates.

At 60 and 45 pages per minute respectively, the DocuColor 2060 and 2045 have the ability to monitor image quality and make required adjustments continuously to produce consistent, high-quality output. DocuColor 40 Pro prints at 40 pages per minute single-sided and 30 pages per minute double-sided. It also features an assortment of digital front ends and accessories.

The DocuPrint 350, 700, and 1000 are continuous feed black-and-white printers that produce 353, 708, and 1,002 images per minute, respectively. The 700 and 1000 models consist of twin print engines that allow duplex printing. The DocuPrint 500 CF is a continuous feed printer for print-on-demand, transactional environments that prints at 230 feet per minute.

The DocuPrint 75 is a cut-sheet, monochrome printer that prints at 75 pages per minute. The DocuPrint 90 is a cut-sheet, monochrome digital printer for small and mid-size print operations. It prints at speeds up to 90 pages per minute and can create on-demand, personalized documents. The DocuPrint 92 C prints at 92 pages per minute and offers black-plus-one-color production printing. The DocuPrint 115 and 155 EPS black-and-white printing systems print at 115 and 155 pages per minute, respectively.

Océ

All four of Océ's DemandStream 8000 DI printers are built from the same print engine with a Scaleable Raster Architecture (SRA) print controller. This common platform allows each member of the family to be upgraded in speed and print volume.

The DemandStream 8090cx features an enhanced processing print controller that is scaleable in memory and processing power to accommodate fast processing of complex pages. The SRA3 controller comes standard with Multiple Resolution Mode for support of 600 dpi resolution output and three interfaces (one TCP/IP interface and two customer-selectable interfaces). An 18.25-inch print width allows two-up 6x9-inch image printing at 1.8 times more images per minute (ipm) than two-up printing.

Océ also offers the VarioPrint 5000 cut-sheet printing systems that offer output speeds ranging from 108 images per minute to 155 images per minute with resolutions up to 600 dpi. It supports a wide range of substrates, including plain white paper stock, custom size formats, coated stock, colored paper, card stock, recycled content paper, preprinted forms, adhesive labels, carbonless forms, pressure seal, perforated, punched or drilled, and tab stock.

Xeikon

Xeikon's DCP 500 D is a 20-inch digital color press that uses the IntelliStream front-end, which accepts the PPML/VDX variable data standard and PPML variable data language. It can print up to 8,280 pages per hour.

The Xeikon 5000 is for short print runs as well as for runs of several million copies. It can process volumes of up to three million A4 color pages a month. Toner can be added during a print run, and press alignment is automated. The press is fitted with a fifth color station and is capable of printing on a wide spread of media weights. The digital front end, the X-800, provides a workflow "toolbox." The controller is able to process jobs incorporating graphics with variable data from over one million records.

Sakurai

Sakurai Graphic Systems incorporated Presstek's ProFire DI technologies into the 29-inch Oliver 474EPII DI press. The press is hybrid, combining direct imaging technology with the capability of running conventional plates. The four-page press is available with a perfector.

The Oliver 474EPII DI is fully automated to accommodate press proofing and short runs. It has automatic plate changing, automatic ink density control, and a workstation that integrates all of the press operations.

9

Post Press and Finishing

Post press and finishing are all of those processes that take place after printing has occurred. The processes include such applications as folding, collating, trimming, stitching, gluing, binding, scoring, embossing, foil stamping, die-cutting, and more. These processes can be inline or offline. Inline means that the processes take place on the printing press but after the ink or other applications, such as varnish or coating, have been applied to the substrate. Offline means that the processes take place on separate equipment after printing is completed.

Printed sheets must often be folded to create signatures having the proper page sequence for a brochure, book, magazine, or other documents that have sequential pages. Many documents require more than one signature that must come together to form the finished multi-page document, such as a book. Bringing together these signatures is called collating and there are machines that do this.

Once collated, the signatures must be trimmed so that the edges are even and the front edge is flat. In fact, nearly all printing requires trimming to some extent. Large hydraulic or electronic paper-cutting machines are used for trimming large stacks of paper.

Bookbinding samples
(Smith Falls Book Binding)

Printed documents having more than four pages must often be stitched or glued, and there are various applications to do this. Stitching can be in the form of saddle stitching with the end result looking like staples in the spine of the document, or stitching can be achieved with thread such as in the case of well-bound books. Gluing also secures the pages in a document. However, glued pages are often not as secure as stitched pages. Stitching and/or gluing are considered the essence of what is called binding.

Let's look at some other finishing processes. Die-cutting is a process where a segment of the printed document is cut into a special shape such as a circle, triangle, or any other shape for that matter. Die-cutting is common in package printing where each final piece has a structure or shape of its own.

Scoring is a process where a printed sheet is given a fine-line indentation to aid the folding process. Similarly, a technique called perforating, as the term implies, creates a line of perforations in the substrate also to aid in the folding process or to make it easier for the user of the printed piece to tear out a section—such as a coupon or mail-in card.

Foil stamping is a process that allows the application of a metallic look to a printed piece. Gold or silver foil is commonly used in this process.

Post press operations consist of four major processes: Cutting, folding, assembling, and binding. Not all printed products, however, are subjected to all of the processes. For example, simple folded brochures do not undergo binding.

There are many additional lesser post press finishing processes such as varnishing, perforating, and drilling. Some types of greeting cards are dusted with gold bronze. Printed metal products are formed into containers of various sizes and shapes. Containers may also be coated on the inside to protect the eventual contents. Other substrates may be subjected to finishing processes that involve pasting, mounting, laminating, and collating. There are also a number of post press operations unique to screen printing including vacuum forming and embossing.

A limited number and volume of chemicals are used in post press operations. The major types of chemicals used in post press are the adhesives used in binding and other assembly operations. The following is a brief overview of each of the four major post press operations.

Cutting

The machine typically used for cutting substrates into individual pages or sheets is called a guillotine cutter or "paper cutter." These machines are built in many sizes, capacities, and configurations. In general, however, the cutter consists of a flat bed or table that holds the stack of paper to be cut. At the rear of the cutter the stack of paper rests against an adjustable back guide. The back guide allows the operator to accurately position the paper for the specified cut. The side guides or walls of the cutter are at exact right angles to the bed. A clamp is lowered into contact with the top of the paper stack to hold the stack in place while it is cut. An electric engine operating a hydraulic pump normally powers the cutting blade itself.

To assist the operator in handling large reams of paper, which can weigh as much as 200 pounds, some tables are designed to blow air through small openings in the bed of the table. The air lifts the stack of paper slightly, providing a near frictionless surface on which to move the paper stack.

The cutter operator uses a cutting layout to guide the cutting operation. Typically, the layout is one sheet from the printing job that has been ruled to show the location and order of the cuts to be made. Though cutting is generally considered a post press operation, most lithographic and gravure web presses have integrated cutters as well as equipment to perform related operations such as slicing and perforating.

Folding

Folding largely completes post press operations for certain products such as simple folded pamphlets. Other products are folded into signatures of four pages or more. Multiple signatures are then assembled and bound into books and magazines. Though folding is generally considered a post press operation, most lithographic and gravure web presses are equipped with folders.

Three different folders are used in modern printing companies. They range in complexity from the bone folder to the buckle folder. Bone folders have been used for centuries and are made of either bone or plastic. These folders are simple shaped pieces of bone or plastic that are passed over the fold to form a sharp crease. Today, they continue to be used, but only for small, very high-quality jobs.

Knife folders use a thin knife to force the paper between two rollers that are counter-rotating. This folds the paper at the point where the knife contacts it. A fold gauge and a moveable side bar are used to position the paper in the machine before the knife forces the paper between the rollers. The rollers have knurled surfaces that grip the paper and crease it. The paper then passes out of the folder and into a gathering station. Paper paths, knives, and roller sets can be stacked to create several folds on the same sheet as it passes from one folding station to another.

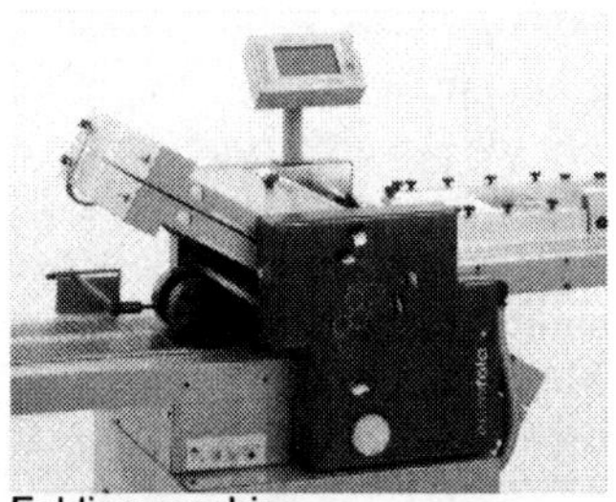

Folding machine
(Eurofold)

Buckle folders differ from knife folders in that the sheet is made to buckle and pass between the two rotating rollers of its own accord. In a buckle folder, drive rollers cause the sheet to pass between a set of closely spaced folding plates. When the sheet comes in contact with the sheet gauge, the drive rollers continue to drive the paper, causing it to buckle over and then pass between the folding rollers.

Assembly
The assembly process brings all of the printed and non-printed elements of the final product together prior to binding. Assembly usually includes three steps: gathering, collating, and inserting. Gathering is the process of placing signatures next to one another. Typically, gathering is used for assembling books that have page thicknesses of at least three-eighths of an inch. Collating is the process of gathering together individual sheets of paper instead of signatures. Inserting is the process of combining signatures by placing or "inserting" one inside another. Inserting is normally used for pieces having a final thickness of less than one-half inch.

Collator, stitcher, trimmer (Muller Martini)

Assembly processes can be manual, semiautomatic, or fully automatic. In manual assembly operations, workers hand-assemble pieces from stacks of sheets or signatures laid out on tables.

Sheets or signatures are picked up from the stacks in the correct order and either gathered, collated, or inserted to form bindery units. Some printers use circular revolving tables to assist in this process. However, due to the high cost of labor, manual assembly is used only for small jobs.

Semi-automatic assembly is completely automated except that stacks of sheets or signatures must be manually loaded into the feeder units. During semi-automatic inserting, operators at each feeder station open signatures and place them at the "saddlebar" on a moving conveyer. The number of signatures in the completed publication determines the number of stations on the machine. Completed units are removed at the end of the conveyer and passed on to the bindery.

Automatic assemblers are similar to semi-automatic units except that a machine and not a person delivers the sheets or signatures to the feeder station and places them on the conveyor. In order to improve efficiency, automatic assemblers are typically placed in line with bindery equipment.

Binding Methods
Binding is categorized by the method used to hold units of printed material together. The three most commonly used methods are adhesive binding, side binding, and saddle binding. Three types of covers are available to complete the binding process: self-covers, soft-covers, and case bound covers.

Adhesive binding, also known as padding, is the simplest form of binding. It is used for notepads and paperback books, among other products. In the adhesive binding process, a pile of paper is clamped securely together in a press. Liquid glue is then applied with a brush to the binding edge. The glue most commonly used in binding is water-soluble latex that becomes impervious to water when it dries.

For notepads, the glue used is flexible and will easily release an individual sheet of paper when the sheet is pulled away from the binding. Adhesive bindings are also used for paperback books, but these bindings must be strong enough to prevent pages from pulling out during normal use. For paperback bookbinding, hot-melt glue with much greater adhesive strength than water-soluble latex is applied. In one version of this method is called perfect binding. In as more permanent process, a piece of gauze-like material is inserted into the glue to provide added strength.

Perfect binding machine
(Muller Martini)

In side binding, a fastening device is passed at a right angle through a pile of paper. Stapling is an example of a simple form of side binding. The three other types of side binding are mechanical, loose-leaf, and side-sewn.

A common example of a form of mechanical binding is the metal spiral notebook. In this method of binding, a series of holes are punched or drilled through the pages and cover and then a wire is run through the holes. Mechanical binding is generally considered as permanent; however, plastic spiral bindings are available that can be removed without either tearing the pages or destroying the binding material. Mechanical binding generally requires some manual labor.

Loose-leaf bindings generally allow for the removal and addition of pages. This type of binding includes the well-known three-ring binder.

Side-sewn binding involves drilling an odd number of holes in the binding edge of the unit and then clamping the unit to prevent it from moving. A needle and thread is then passed through each hole proceeding from one end of the book to the other and then back again to the beginning point. This type of stitch is called a buck-stitch. The thread is tied off to finish the process. Both semi-automatic and automatic machines are widely used to perform side stitching. The main disadvantage of this type of binding is that the book will not lie flat when opened.

In saddle binding one or more signatures are fastened along their folded edge. The term saddle binding comes from an open signature's resemblance to an inverted riding saddle. Saddle binding is used extensively for news magazines where wire stitches are placed in the fold of the signatures. Most saddle stitching is performed automatically in-line during post press operations. Large manually operated staplers are used for small printing jobs.

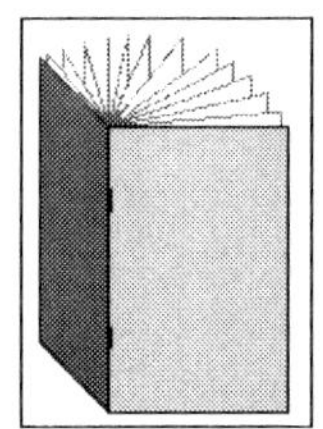
Saddle stitching (Prestoprint)

Another saddle binding process called Smythe sewing is a center sewing process. It is considered to be the highest quality fastening method used today and will produce a book that will lie almost flat.

Covers

Self-covers are made from the same material as the body of the printed product. Newspapers are a common example of a printed product that uses self-covers. Soft covers are made from paper or paper fiber material that is somewhat heavier or more substantial than the paper used for the body of the publication. This type of cover provides minimal protection for the contents. Unlike self-cover, soft covers almost never contain part of the text of the publication. A typical example of the soft cover is found on paperback books. These covers are usually cut flush with the inside pages and are attached to the signatures by glue (though they can also be sewn in place).

Case-bound covers are the rigid covers generally associated with high-quality bound books. This method of covering is considerably more complicated and expensive than any of the other methods. Signatures are trimmed by three-knife trimming machines to produce three different lengths of signatures. This forms a rounded front (open) edge to give the finished book an attractive appearance and provides a back edge shape that is compatible with the shape of the cover. A backing is applied by clamping the book in place and splaying or mushrooming out the fastened edges of the signatures. This makes the rounding operation permanent and produces a ridge for the case-bound cover.

Gauze and strips of paper are then glued to the back edge in a process called lining up. The gauze is known as "crash" and the paper strips are called "backing paper." These parts are eventually glued to the case for improved strength and stability. Headbands are applied to the head and tail of the book for decorative purposes. The case is made of two pieces of thick board, called binder's board, that is glued to the covering cloth or leather. The covering material can be printed either before or after gluing by hot-stamping or screen methods. The final step in case binding consists of applying end sheets to attach the case to the body of the book.

Inline Finishing

Historically, the finishing operations just described were labor-intensive operations handled within printing company or by a trade bindery. Even when performed within a printing company, finishing operations generally were not integrated with the presses or with each other. Today web presses are often linked directly to computer-controlled inline finishing equipment. Equipment is available to perform virtually all major post-press operations including cutting, folding, perforating, trimming, and stitching. Inline finishing equipment can also be used to prepare materials for mailing. The computer can store and provide addresses to inkjet or label printers, which then address each publication in order of zip code.

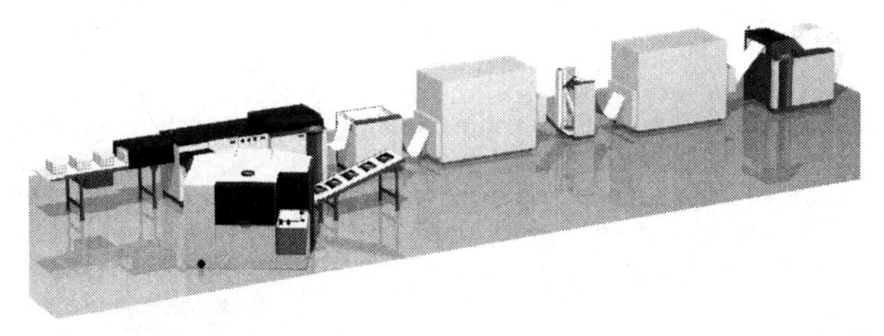

Inline binding system
(Muller Martini)

One of the most important results of computer inline finishing is the introduction of demographic binding, the selective assembly of a publication based on any one or more of a number of considerations including geographic area, family structure, income, or interests. For example, an advertisement can appear only in those copies of a magazine intended for distribution in an advertiser's selling area. Demographic binding has proven to be a successful marketing tool and is already widely used, especially by major magazines.

Inline finishing equipment can reduce the number of operators and helpers required for offline finishing operations while increasing the rate of production.

Binding and Finishing Considerations for Digital Printing

With digital printing growing in popularity, many service providers are finding they now need to offer a variety of binding and finishing services. Commercial printers, publishers, direct marketers, and service bureaus can offer digital print services: shorter print runs; delivery on demand; variable, customized, one-to-one content; as well as high-quality color. These are all value-added services. However, these services mean little if the binding and finishing applied to the printed products are not high quality in appearance.

Traditional binderies designed to work with offset presses may not have a range of digital finishing capabilities suited for today's digital press output. Therefore, printers deciding to bring the finishing process into the printing company may need to equip their plants with binderies geared more specifically to work with their digital presses.

Features of a "Digital Bindery"

Digital post press has typically lagged behind developments in prepress and press technology. However, more recently key digital press manufacturers such as Xerox, IBM, Océ, HP Indigo, Xeikon, and Kodak NexPress have been working with finishing vendors in original equipment manufacturers (OEM) and partner roles to help provide printers with bindery solutions optimized for use with roll-fed and cut-sheet digital presses. However, the big challenges are developing better binding of digital color output, bringing the job definition format (JDF) job ticket to more inline finishing devices, and adoption of standardized communications interfaces linking post press to press and prepress processes.

Binding Digital Output

There are several critical questions to explore in order to understand the relationship between digital printing and binding and finishing options. For example: What does it mean for a bindery to be "optimized for a digital press"? Why can't printers equipped with offset binderies use them to finish their digitally printed products? How does a "digital bindery" differ from a traditional one?

Simpler digital print jobs can use some traditional binderies for finishing. However, a number of digital applications will find these binderies inefficient. To begin with, traditional bindery equipment for offset presses are typically mechanical devices, designed to handle medium to long runs of identically finished jobs, which must be collated in a separate step before they are bound. By contrast, binderies optimized for digital presses must be able to handle very short runs, even a single copy, of full-page sets, as well as variable output without incurring production bottlenecks or requiring excessive makeready.

Job control is also important. Optimally as part of a JDF workflow, job parameters are specified up front in the job ticket but can shift rapidly for variable data printing. Also, control must extend to the equipment involved and to intelligence about the state of the product in the binding process itself—such as page order and number, insertions, covers—while handling errors that may occur. This can be done through direct feedback within inline solutions, or by using bar codes or related marks.

Next, because digital presses print using toner or electrostatically applied inks rather than offset inks, they have special handling requirements that must be addressed. Because toner rests on the surface of the paper (unlike offset ink, which is absorbed by the paper fibers), toner-based print is more prone to scratching and rub-off from finishing processes such as cutting, folding, and booklet making. Thus, vacuum rather than friction feed should be part of the bindery. Additionally, due to the chemical properties and high heat surrounding electrostatically applied inks, color pages are susceptible to cracking if scoring is not done properly or if the direction of paper grain when creasing and folding is not taken into account.

Other idiosyncrasies of digitally printed pages include more image drift than with offset (as much as a 1.5 mm variance in image position in any direction) and, thus, slightly skewed registration from front to back, requiring larger bleeds. Sophisticated binderies will include cutters that can read the actual position of the image on the page and adjust accordingly. Similarly, digital printers have a tendency to leave residues of silicon on the page. This residue can cause poor glue adherence when gluing the spine of a book block. Therefore, adequate milling and notching tools must be an intrinsic part of a digital perfect binding process. Finally, because digital pages can cling to each other because of static in a sheet-fed bindery, an antistatic function in the bindery feeder will be critical to ensure the pages are fed one by one.

To optimize a digital workflow, digital binderies often need to be set up inline or nearline. An inline bindery is tightly coupled to the press, while a nearline bindery is connected over a network (or can read marks or bar codes for details of the job). Offline setup also is possible, where the binder is completely separate from the press, allowing multiple press feeds (including from offset presses). However, the chain is broken for true digital control. One solution to this problem is a hybrid nearline device that can work in a mixed environment where collating towers can collate sets of offset printed sheets that have been loaded alongside a sheet-feeding device that takes sequentially printed sheet sets typical of that from digital presses.

Most sophisticated binderies work in tandem with other modules from the same or different vendors, including feeders (both roll-fed and sheet-fed), rewinders (for roll-fed), creasers, trimmers, and collating towers. For perfect binders, they work with cooling towers for hot-glue solutions, as well as laminating devices for cover finishing. Even stackers and packagers may be part of the equipment mix.

Similar to offset binderies, digital bindery solutions are generally optimized to work with roll-fed or cut-sheet presses, as with black-and-white or color output, though bringing monochrome and color pages together is often a key part of the bindery process. Electronic binderies also span a range of types from loose bound (spiral, wire, plastic, and ring) to stitched (thread or staple saddle stitching-centered on the spine, or alternately, side or corner-stitching), extending to perfect-bound and hardcover. Saddle-stitch and perfect-bound binderies are the primary focus of digital finishing vendors today, though "digital book factories" may also include solutions for hardbound books.

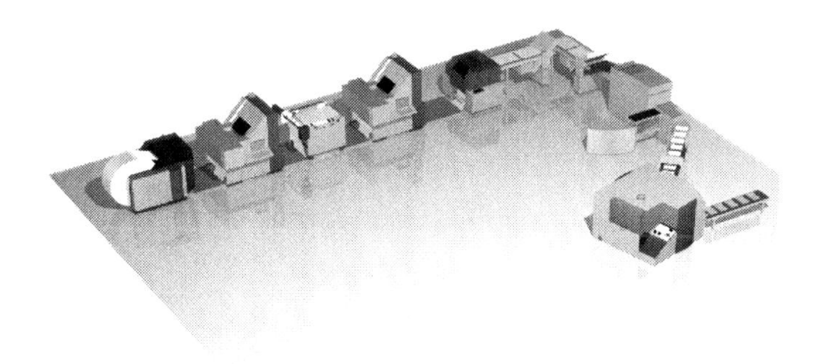

Inline digital binding workflow
(Muller Martini and Nipson)

The Future

As digital print suppliers approach their digital press and finishing vendors for help in deciding which bindery to use, uppermost in their minds should be the kind of press they have, the job they do, and the level of communication they require between press and bindery. Many of the digital press manufacturers and bindery vendors have varying degrees of "digital optimization." Moreover, as the JDF and Universal Printer, Pre- and Post-Processing Interface (UP3i) standards advance into adoption, upgrades to vendors' finishing systems to incorporate these standards will be an important consideration.

The Universal Printer, Pre- and Post-Processing Interface (UP3i)

When digital print suppliers are looking for finishing devices that will work with their workflows, they are faced with the same questions regarding the finishing vendors. For example: How do you communicate between the digital devices? Until now, digital press vendors have provided proprietary answers, using their own control languages. When mixing and matching, however, it is too expensive for finishing vendors to be writing new code for every digital press to which they may want to connect.

Over the past few years, with the growing demand for integrated (inline) or semi-integrated (nearline) finishing, it became clear that a new open communications standard was needed to allow digital presses and finishing devices to talk to each other. Therefore, a group consisting of Océ, Xerox, IBM, Stralfors, Hunkeler, and Duplo formed to develop the UP3i. UP3i is designed to promote standardized, end-to-end communication of systems data between the printer and the pre- and post-processing hardware of an integrated digital print line. UP3i has various aims including:

- To shorten setup and preparation times
- To enable remote setup
- To maximize the efficiency of any printing or finishing line
- To facilitate implementation with job tickets (especially JDF)
- To support both continuous form and cut-sheet printers
- To enhance job recovery in the case of paper jams

UP3i is compatible with and complementary to the JDF standard developed by CIP4. CIP4 is an international, worldwide operating standards body located in Switzerland. The purpose of the association is to encourage computer-based integration of all processes that have to be considered in the graphic arts industry—in particular the specification of standards, such as the new Job Definition Format. JDF defines the job ticket, while UP3i defines the interface between the devices. Thus, a consideration for digital print suppliers who are planning to invest in new digital finishing devices should be whether the vendor is compliant—or will be—with UP3i. As printing companies upgrade their equipment, UP3i will become increasingly important.

10

DRUPA

DRUPA is a German acronym standing for *druck* (printing) and *papier* (paper). DRUPA is the world's largest graphic communication exposition that takes place every four to five years in Dusseldorf, Germany. It originated in 1951 and is attended by approximately a half million people from all over the world. Approximately 2,000 exhibitors of technology and related items show more than 100,000 products over a three-week period.

DRUPA defines the direction of the graphic communication industry. This chapter provides a summary of industry trends defined by four DRUPAs: 1990, 1995, 2000, and 2004.

DRUPA 1990

DRUPA 1990 represented the "digital divide." It showed that the direction of the industry was digital and that analog technology would greatly diminish in the years to follow.

Prepress

Prepress at DRUPA 1990 exhibited the first computer-driven workstations and greater computerized production for the graphic communication industry. There was a search in the industry for higher quality and productivity, and the role of designer and color technician began to merge with the introduction of low-cost desktop publishing systems (under $10,000) as mid-range color systems ($10,000 to under $50,000) started to replace high-end systems ($50,000 to over $1 million) of the 1980s.

DRUPA Logo

Adobe PostScript and Apple Computer revolutionized prepress and showed how a major integration of digital technology would allow prepress production at greater speeds and lower costs than ever before. PostScript became the default page description language to drive graphic arts technology, particularly at the front-end. It drove imagesetters and workstations of all kinds for color and non-color systems, as well as systems that produced text and full-color pages of all sizes for commercial and publication printing. PostScript's interpreter software enabled its use by Hell, Crosfield, Scitex, and Dianippon for their popular integrated scanner systems of the time. It allowed the development of modular color separation systems in which companies could mix and match components from different prepress vendors and not be locked into using one vendor for nearly everything. Because of the role of Adobe PostScript and Apple Computer, the industry began to experience the production of low-cost color pages.

DRUPA 1990 showed products resulting from the trend toward mergers, joint ventures, and acquisitions in the graphic communication industry. To improve the technology of color proofing, Stork participated in a joint venture with Scitex, as did Kodak, and 3M worked with Apple to use Macintosh computers for Matchprint digital proofing.

The trend toward increased mergers, joint ventures, and acquisitions was to continue into the twenty-first century.

DRUPA 1990 showed the graphic communication industry adopting technologies related to artificial intelligence (AI), direct color proofing using no film, optical disk systems, high definition television (HDTV) for viewing images for catalog printing, the use of Macintosh computers for pagination and color systems. DRUPA 1990 predicted the beginning of the end for high-cost color separation systems.

Press

In the area of press, DRUPA 1990 showed the first computer keyboard controlled press systems and press speeds of 40,000, 50,000, and 60,000 impressions per hour. It also showed web printing inspection by computer chip video cameras and sensors for electronic ink and water balance on offset lithographic presses. Robot controls were introduced in some areas of printing.

Approximately 450,000 people from around The world attends DRUPA.

Color measurement and control systems were developed for monitoring this high-speed printing press output and DRUPA 1990 showed a proliferation of color control devices for colorimetry using instruments that “saw” color the way the human eye does.

There were demonstrations of flexography using water-based inks, which shortly afterwards was adopted by some newspaper printers to solve the ink rub-off problem of standard oil-based inks.

In sum, DRUPA 1990 showed that the printing press had been greatly advanced through electronic controls to run faster and produce higher and more consistent quality through improved control systems.

Post Press

While most of the focus of DRUPA 1990 was on development in prepress and to a lesser extent on press, there were also some major developments in the area of post press or binding and finishing. It was evident that in the coming years post-press technology was going to introduce computer-driven controls and automation through the use of microprocessors. There was promise that the entire post press area was going to be revolutionized from a systems engineering point of view, and this indeed happened.

The prospects for the future of the graphic communication industry following DRUPA 1990 were promising. Due to computers, digital technology, and electronic controls, the cost of production was going down while productivity was increasing. Low-cost workstations were replacing high-cost color systems. An integration of bindery into press was imminent, and the promise of a "filmless," "pressless" color publishing system was real.

DRUPA 1995

DRUPA 1995 showed the graphic communication industry transforming into a global communication industry in which information for printing and publishing was produced and transmitted any place in the world at faster speeds than ever before.

The clear themes of DRUPA 1995 were new evolution of past systems and revolutions in new ways to digitally prepare and transmit information. Digital imaging dominated DRUPA 1995 with digital cameras, digital scanners, digital prepress, digital management systems, digital printing, and the digital bindery.

"Print City" is a main DRUPA attraction.

Electronic data exchange using common languages became standard for the transmission of information to and from remote sites. ISDN (Integrated System Digital Network) showed great promise for the industry as the need for enhanced speed became apparent. DRUPA 1995 also showed faster RIPs, automated make-ready, and the "green" revolution, including dry film requiring no chemical processing as well as water-based chemistry.

There was the further proliferation of systems allowing higher productivity at less cost. Strategic positioning and partnering continued through alliances between former competitors such as joint ventures and partnerships. In 1995 there was enhanced participation in industry associations and consortia by equipment manufacturers in order to share information and develop. It was evident that the prevailing attitude had become that no one company could solve all of the problems or address all of the needs of the industry and that vendors had to work together and share knowledge.

Prepress

DRUPA 1995 showed prepress becoming the gateway to press. In other words, prepress and press systems were going to be integrated. Prepress was going to "drive" press, and prepress companies of the future would be data-based management companies involved in moving data from prepress to press plates or press cylinders. This would be done through integration of systems, software, and telecommunication.

Adobe Systems set the standards for much of the software driving prepress operations with page management systems and cross-platforming open systems. However, in 1995, Macintosh computers represented over 90 percent of the workstations used in the graphic communication industry.

Direct-to-plate systems were numerous for metal and polyester plates at DRUPA 1995. However, there was evidence that proofs would be around for a long time even though prepress was moving from the prepress vendor or "trade shop" to the printer. With printers then getting involved in their own prepress using desktop systems, prepress vendors started looking for new ways to expand their services. Some started "pressing" CDs as well as providing interactive media services. It became evident at DRUPA 1995 that for traditional graphic communication companies to remain viable, they needed to diversify into new products and services and redefine themselves as "communication companies," not merely printing companies.

It was clear, however, that paper will be used for a long time to come, though in a more strategic way than in the past. There were signs in 1995 of an industry move to variable data printing and targeted messages, and short-run color printing as opposed to the typical long run color printing that defined a large portion of the graphic communication industry for nearly a century. The move in this direction was to be driven by output devices. New prepress technology was capable of producing plate-ready film with no image assembly and everything done from the workstation including color measurement, the production of spectrally based products, and desktop color management of prepress and press.

Press

DRUPA 1995 showed new digital and automated capabilities of printing presses. Fewer but larger press units were being sold for high-volume printing, but more color printing was shifting to short runs on digital presses. Some of these presses required no film, no plates, and no make-ready. Indigo's E-Print 1000 was the first such press, and many others followed shortly thereafter. Advances in printing press technology in 1995 were driven by the market's affinity to paper having the benefits of consumer convenience, privacy, high quality, and portability. This affinity was evident in spite of rising paper costs.

Post Press

Post press advances at DRUPA 1995 showed continued automation, much faster make-ready, and overall faster systems.

In sum, whereas DRUPA 1990 had its main focus on prepress, DRUPA 1995 showed a shift to an emphasis on press, and particularly digital presses.

DRUPA 2000

DRUPA 2000 addressed the question: What new technologies will drive the graphic communication industry segments into the future?

DRUPA 90 and 95 were evolutionary and revolutionary with their introduction and development of digital technologies including digital project management, digital printing, digital press controls, and the digital bindery. These two previous DRUPA expositions showed that the differences between prepress and press were blurring, and DRUPA 2000 showed the continuation of digital developments with technologies becoming interlaced and interactive. It was a show of the "cyberintegration" that demonstrated the potential of the graphic communication industry in using online technology for imaging production and distribution.

Prepress

The prepress areas that were to define the graphic communication industry of the next four to five years included more CD-ROM and DVD use; media-independent workflows; image management data-based systems; prepress involving information and image management; value added services; high-quality on-demand printing; a variety of multi-media services; more software tools and integration of software; more moving, storing, and manipulation of data; enhanced application of computer-to-plate systems; more digital proofing; and digital proofs requiring color management.

Whereas DRUPA 95 introduced the "Green Revolution" in graphic communication, DRUPA 2000 showed the real application and adoption of it across the industry with the greatly enhanced use of dry film and water-based chemicals and a significant focus on less waste and appropriate disposal of waste products. There was the promise of even more digital and environmentally friendly prepress technologies developing in the years ahead.

DRUPA 2000 raised many questions about what prepress had come to mean and indicated that prepress, as it has been known in the past, may have become obsolete.

Press

In the area of printing presses, DRUPA 95 showed advances in digital and automated press technology. Questions on everyone's mind were: What is the future of printing technology? Will it be replaced with the Internet and World Wide Web, video, holography, or an as-yet-to-be-developed medium?

DRUPA 2000 showed that planning and development in printing press technology were beginning to specifically address customer needs, particularly in the area of on-demand printing, short-run color, and variable data printing. There was more computer-integrated manufacturing (CIM) and CIP-compliant technology showing an integration of prepress, press, and post press—the processes were being tied together. There was promise that CIP III and computer-integrated manufacturing would grow rapidly in integrating the printing press with prepress and post press technologies.

CIP is an international, worldwide operating standards body located in Switzerland. The purpose of the association is to encourage computer-based integration of all processes that have to be considered in the graphic arts industry, in particular the specification of standards.

Project storage to file management, more typically associated with prepress, had found its way to the printing press at DRUPA 2000, as did two categories of production systems for image distribution. They were printing press cylinders that were re-imaging for each page, and as well as the traditional presses that use a fixed image on a plate to produce the same image for every page. However, printing presses that allowed the production of variable data personalized pages predominated new press developments with their focus on higher speeds, short-run color, and more digital printing overall. This was a defining development that clearly showed the direction that printing presses would take.

Post Press

Past DRUPA expositions showed modest developments in finishing automation. Binding and finishing technology, for the most part, had lagged behind developments in prepress and press. However, this was about to change with continued focus on CIP standards, inline press to finishing operations, and the need for post press operation speeds that could keep up with the increasing speeds of presses.

DRUPA 2000 provided a glimpse at the developments that would drive the bindery in the twenty-first century. DRUPA 2000 showed significant amounts of automation in moving paper—e.g., from production of printing to delivering finished printing jobs to loading docks. It showed laser control using bar coding, computer-controlled finishing systems, new levels of automation for setting up finishing equipment, automation to remove jobs from press and move them to the bindery, automation for bundling and wrapping, and even post press equipment that exceeded press speeds.

Much of what was shown at DRUPA 2000 in binding and finishing were improvements to existing technology to do more and to do it better. However, there were post press technologies shown that had never been seen before.

DRUPA 2004

DRUPA 2004 showed an enormous array of the graphic communication profession's present-state of technology and visions for the future. A central theme was "hybrid technology" with different technologies borrowing from each other to create new and improved technologies. This was a synergy that is likely to continue.

DRUPA 2004 was also the first "filmless" DRUPA. No film products were shown, and there were very few scanner exhibits because of this. However, digital printing, inkjet, wide format printers, and digital photography were dominant. Trends in these areas included more vibrant images, larger varieties, larger formats, faster speeds, and more durability with substrates, inks, and toners. The major press manufacturers—Heidelberg and MAN Roland—showed sheet-fed presses having speeds of 18,000 IPH that were virtually noiseless.

If DRUPA 2000 was characterized as a show of the potential of "cyberintegration," DRUPA 2004 was "Cyberintegration 2." JDF (Job Definition Format) was everywhere and interconnection was a main theme. JDF is a comprehensive XML-based file format/proposed industry standard for end-to-end job ticket specifications combined with a message description standard and message interchange protocol. JDF is designed to streamline information exchange between different applications and systems. It is intended to enable the entire industry—including media, design, graphic arts, on-demand, and e-commerce companies—to implement and work with individual workflow solutions. JDF allows integration of heterogeneous products from diverse vendors to seamless workflow solutions.

XML stands for Extensible Markup Language. It is a simple, very flexible text format derived from SGML. Originally designed to meet the challenges of large-scale electronic publishing, XML is also playing an increasingly important role in the exchange of a wide variety of data on the Web and elsewhere. SGML, Standard Generalized Markup Language, is an enabling technology used in applications such as HTML. HTML is the basis of the language use for developing World Wide Web pages.

While there were no particularly new major developments in the form of revolutionary products, there were many upgrades and improvements to existing technology and software. JMF (Job Messaging Format), a component of JDF, was seen providing data collection and queries for feedback to downstream processes. JMF closes the loop for true Computer Integrated Manufacturing (CIM). There were vast improvements in digital presses and in variable data printing software, as personalized printing has become a high growth area of graphic communication. Additionally, new finishing techniques were shown, particularly in the areas of inline finishing for high-quality packaging printing. Flexographic units for inline finishing and sheetfed offset printing were ever-present as was UV technology in sheetfed offset with multiple finishing effects.

The predominant companies that displayed at DRUPA 2004 defined the direction of graphic communication technology. The larger exhibitors in prepress were Agfa and Kodak Polychrome Graphics. In traditional presses they were Goss, KBA, Heidelberg, and MAN Roland. In digital presses they were Hewlett Packard, Océ, Scitex, Screen, Xeikon, and Xerox. In post press it was Muller Martini.

Incremental advances were the characterization most frequently applied to developments shown at DRUPA 2004. Instead of mega-trends, attendees saw refinements in selling propositions and convergence of market forces. Companies that previously had invested only in traditional technology were now looking seriously at the opportunities that digital printing offered.

An interesting observation was that monochromatic printing was experiencing a revival in the face of all of the emphasis on short-run color. Several manufacturers launched new monochrome printing solutions. This came at a time when more users of digital color devices were moving toward running all jobs in color, black and white, or hybrid on a single production machine for processing efficiency. Higher volumes still dictated the use of dedicated black-and-white high-speed digital presses.

Prepress

In prepress there was a vast showing of computer-to-plate (CTP) devices with laser, UV, and thermal exposure capabilities. Creo introduced a waterless polyester plate for direct-imaging presses and a processless aluminum plate. The polyester plate is suitable for run lengths of up to 30,000 impressions.

In a technology demonstration, a Creo Trendsetter 800 Quantum CTP platesetter imaged a processless plate. The plate required no gumming, processing, or post-imaging treatment. It was designed for run lengths up to 50,000 impressions. Creo also launched a new Spire color server to drive a Xerox DocuColor color printer.

ECRM showed the MAKO four-up CTP platesetter that can image a six-page gatefold with register marks and color bars in position. Plates in a cassette were automatically fed into the integrated registration system.

EFI introduced version 2.0 of its OneFlow PDF-based prepress workflow software. Designed to streamline two- and four-up digital and offset printing and proofing, the software's new features included integrated proofing and imposition, a hybrid screening tool, interactive workflow management, browser-based job tracking, and CIP4-compliant JDF connectivity. CIP4 represents standards for digital integration of prepress, press, and post press operations.

Esko-Graphics introduced Scope workflow software and hardware components that cover a range of functions, from specification, design, and pre-production operations to platemaking (for printing) and tool making (for converting). It also has capabilities for project coordination, digital asset management, and distributed proofing and approval. Esko-Graphics also showed a four-up CTP system that images conventional presensitized plates with UV light.

Glunz & Jensen introduced the PlateWriter 4200, which produces press-ready aluminum plates without chemical processing for small- and medium-format presses. The PlateWriter system consists of the PlateWriter imaging engine, a finishing unit with a gumming station, and a hardware or software RIP, including a Harlequin-based Xitron software RIP. The device jets a patented Liquid Dot solution onto non-photosensitive two- and four-up plates. The imaged plates then are fed through a plate-finishing unit that dries the plates and bonds the "liquid dots" to the plate surface.

Heidelberg presented the Suprasetter series of four- and eight-page, thermal, external-drum CTP platesetters. With a new Heidelberg-developed laser system, the modular series is available in five different speeds, in various automation settings, and with a single or a multi (four) cassette loader, which holds 600 plates.

Kodak Polychrome Graphics (KPG) launched Matchprint ProofPro 4P printing system, a new version of its SWOP-certified Matchprint InkJet proofing system. The RIP is based on Adobe's CPSI technology. The 24-inch network-ready printer is based on Hewlett Packard (HP) technology and features a four-picoliter drop size, six-color printing, and quality modes from 600x600 dpi to 2,400x1,200 dpi. The company says the dye-based inks resist fading and provide rich, deep, and accurate colors and good color stability for proofs.

Agfa's Acento four-up thermal CTP platesetter was available in a variety of configurations, each featuring different levels of automation and speed. The Acento E can produce 10 plates per hour; while the Acento S produces 20 plates per hour. Three plate loaders were available.

Presstek Inc. introduced the new ProFire Excel imaging system and ProFire Digital Media, which work in combination to generate a 16-micron spot that enables 300-lpi printing with support for FM/stochastic and hybrid screening options.

Creo Inc. introduced a new Clarus WL waterless polyester plate material that can be used as a drop-in solution for direct imaging presses. It supports run lengths of up to 30,000 impressions.

There were some new twists thrown in at Drupa 2004, however, including a trend toward integration of offset and digital printing workflows. A case in point was Creo teaming up with Xerox to demonstrate how its Prinergy workflow system could launch Xerox FreeFlow Print Manager and submit jobs for digital printing to Spire color servers driving Xerox digital presses.

HP announced that it had formed an alliance with Quark Inc. to facilitate the production of customized marketing communications. As a result, Quark released a free XTension, called QuarkXClusive, for QuarkXPress 6.0 users that adds variable data publishing capabilities optimized for HP output devices.

Press

Traditional presses at DRUPA 2004 showed improvements in speed and electronic controls. However, digital presses dominated the printing press displays.

Delphax introduced a continuous feed digital web press that prints black-and-white pages at 450 ft/min. At this speed, it can produce nearly 2,000 8.5x11-inch ppm with 600x600 dpi quality on a range of substrates from very lightweight to heavy card stocks.

Heidelberg showed its Speedmaster XL 105 sheetfed press that can print 18,000 29.13x41.33-inch sheets per hour. It is designed specifically for larger-format packaging and label print applications.

Hewlett Packard demonstrated its new seven-color HP Indigo 5000 press and the HP Indigo 3050 press. The HP 5000 can print up to 66 four-color 8.5x11-inch ppm (two-up) or 133 two- or single-color 8.5x11-inch ppm (two-up). The press provides electronic collation and automatic duplexing, and has three software-selectable paper trays. It can produce full-color variable-data printing.

The HP 3050 can print up to 66 four-color, 8.5x11-inch ppm (two-up) or up to 266 8.5x11-inch ppm (two-up). It produces prints with 800x800 dpi resolution and 180 lpi screens. It features the same finishing capabilities as the HP 5000, and it too is a variable data press. The HP Indigo 5000 press is distinctive because it represents the first press co-developed by HP and the former Indigo

Some of the latest technology in graphic communication is displayed at DRUPA. (KBA)

organization, with the collaboration bringing innovations in paper handling, inking systems, and production software.

KBA's Rapida 74 G made is a 29-inch, 15,000 sheet-per-minute waterless offset press featuring the same Gravuflow keyless inking system as is on the 74 Karat DI press. It has a temperature control system, automatic plate changing, remote format and register control, and JDF workflow integration. The press can be configured with up to eight printing units plus perfectors, coaters, and other inline finishing options. KBA's 41-inch Rapida 105 has a high-line delivery extension and can print 18,000 sheets per minute in a straight printing mode or 15,000 sheets per hour in perfecting mode.

Kodak introduced the V series Versamark that has continuous inkjet, high-speed printers. The four-color Versamark V series has a print resolution comparable to 300x1,200 dpi and operates at up to 325 fpm.

Muller Martini demonstrated its SigmaLine, which is an integrated industrial book on-demand system. It is a roll-fed digital press, capable of applying one color to both sides of the paper simultaneously. The press's output is folded inline, then sent to either an integrated perfect binder or an inline saddlestitcher, depending on s job's requirements. The press also has an inline trimmer capable of one-off variable book trimming.

Nipson unveiled the VaryPress 200 and the VaryPress 400. Both monochrome presses produce 600 dpi output at 230 and 410 ft/min. respectively. Both use Nipson's magnetographic imaging and cold flash fusing.

Two new continuous-feed printers were shown by Nipson: the VaryPress 200 and 400, which are based on magnetographic imaging with cold flash fusing. Among the advances incorporated into the printing systems are a new generation of writing heads offering longer life and a smaller spot size, combined with a new toner formulation with a smaller particle size and lower consumption.

Océ demonstrated a VarioStream 7650 printing system coupled with a modular Quick Change Developer Station (QCDS) that prints 1,273 MICR images per minute.

Riso Inc. showed its HC5000 full-color inkjet printer that can print up to 105 letter-size ppm at 600 dpi (900 dpi in fine mode) and can accommodate paper stocks from 14 lb. bond to 60 lb. cover in sizes up to 11x17 inches. The system is suited for volumes from 15,000 to 250,000 copies per month. The printer features a controller and a scanner. Finishing options include punching, stapling, and booklet making.

Xerox demonstrated its continuous-feed DocuPrint 1050 and 525 digital black-and-white printers that can print up to 244 ft/min. at 600 dpi. These printers have a 19.5-inch web width and can accommodate three 6x9-inch images across. The DocuPrint 1050 pairs two DocuPrint 525s to print up to 1,064 two-side, two-up ppm or up to 1,952 two-side, three-up ppm.

Xerox' Nuvera 100 and 120 digital black-and-white printers are engineered to produce offset quality at 100 and 120 ppm, respectively. The printers are available as print-only devices, or with the choice of an integrated 600 dpi, 120-ppm scanner that can scan front and back sides of sheets. It has four standard paper trays, four optional trays, and an optional four-tray insertion module to provide for a total capacity of 17,280 sheets. Print resolution is 600x4,800 dpi and halftone screens range from 85 to 156 lpi. Finishing options include stacking, stapling, booklet making, hole punching, and thermal binding.

Xerox also previewed an optional inline UV coater for the iGen3 and a new version of its digital workflow.

NexPress showed its new Digimaster E150 print system, a 150-ppm, black-and-white press that features enhanced image quality, a modular design, and optional MICR capabilities. In addition, it supports a greater range of substrate weights and sizes (7x8.27 inches to 14.33x18.5 inches) than previous models. The Digital Print Quality Adjust feature enables operators to vary line width and solid area density to achieve a desired printed look.

Delphax Technologies introduced its continuous feed CR2000 monochrome printer. It is capable of producing nearly 2,000 8.5x11-inch ppm using electron beam imaging.

IBM updated its Infoprint 4100 line of continuous-form printers to provide increased paper weight, high-speed (up to 1,220 ipm) MICR capability, and a productivity tracking feature.

Xeikon showed its new Xeikon 5000 digital color press. It offers improved performance with regard to image quality, reliability, and durability. It has a new digital front end that provides greater processing power for production of complex variable data print jobs, even in long runs.

Screen (USA) showed its new TruePress 344. This is a digital offset press that uses processless plate technology from Konica-Minolta to produce up to 7,000 sheets per hour. It handles a maximum sheet size of 13.39x18.5 inches and offers a 2,400 dpi resolution.

MAN Roland showed its digital DICOweb and announced that it is moving into the serial release phase of this press following two years of pilot testing. MAN Roland has incorporated DICO (Digital Change Over) thermal transfer technology for imaging and de-imaging cylinders on-press into heatset and coldest press configurations.

In the area of color measurement for press output, GretagMacbeth introduced the DensiEye 700 four-channel densitometer for CMYK measurement. The company says the unit provides the highest measurement speed and response time on the market. With its optional small aperture and polarization filter, the densitometer is suitable for measuring small color bars. The device includes a set of functionality operators to measure density, dot gain, dot area, trapping, gray balance, and print characteristics.

Other trends at DRUPA 2004 were the applications and opportunities in transaction printing from virtually every digital press manufacturer. This included suppliers of high-end color printing systems that previously had been focused on "graphic arts" applications. Transactional printing is printing of documents used in doing business including contracts, invoices, and related items that can now include variable data printing of advertising and messages to the recipient.

Related to press are inks and toners. BASF showed its NovaArt F 2008 ink series that is designed to widen a sheetfed press's color gamut with enhanced gloss and brilliance as well as print sharpness. Additionally, BASF introduced its Novaspace F 2010 process inks and associated Hyper-space software. These products have made possible a 30 percent greater color space. The company says the ink features pure pigments and produces high optical densities, pure shades, and high brilliance.

Pantone and Xerox announced results of a joint venture that is designed to predict color reproduction from digital printers and presses. The Pantone digital chip system includes 1,089 process color equivalents and provides examples of how these colors reproduce on a Xerox DocuColor 6060 with a Xerox DocuSP front end. According to Pantone, it enables a three-way comparison of solid Pantone colors to the CMYK process-color offset printing simulation and the digital press simulation to provide an accurate visualization prior to printing. All colors are presented in a tear-out chip format on coated stock that can be attached to artwork. The system is designed to allow users to predict how Pantone colors will reproduce on a digital press and to demonstrate which Pantone colors can be reproduced accurately in CMYK.

Post Press

Post press technology at DRUPA 2004 showed systems having higher speeds than previously available resulting from enhanced electronic controls and improved engineering. The goal of this technology was to have post press speeds equal or approach press speeds to enable a continuous flow of production without the occurrence of bottlenecks. Inline finishing systems were also featured by many post press equipment manufacturers. This enabled a continuous production from printing to many facets of binding and finishing. Automated make-ready systems for storing and setting production positions of finishing lines were shown, as were systems that drilled and stitched printed sheets and signatures at speeds never before possible. Inline finishing systems were particularly evident on digital presses.

DRUPA in the Future

Future DRUPA expositions will continue featuring the graphic communication industry at its best. The move to "digital" in nearly all phases of the industry will no longer be considered a revolution, but will be the norm. Likely features will include the integration and connection of digital technologies along many facets of communication, including printing, voice, motion, and the Internet. The power and efficiency of "wireless" technology will have developed and matured by the next DRUPA, and we are likely to see graphic communication technology in the forefront of such advanced methods of sending, receiving, and producing information.

11

Telecommunication

The graphic communication industry's access to broadband communication stems from earlier developments in cable communication and then satellite transmission. The signs were clear that the graphic arts community was going to be favorably impacted by these technologies when their applications were applied to enhanced communication and pictorial image quality of broadcast television.

To understand how the graphic communication industry got to where it is today, one must first understand the historical developments behind cable and satellite communication. The history represents a lesson in how technologies evolve and change to serve the broad needs of society.

Cable Communication

The first cable television system was installed in Astoria, Oregon, in 1948 and followed shortly thereafter with an installation in Lansfield, Pennsylvania, in 1950. The purpose of these early systems was to bring television signals to households in low-lying valleys that could not receive broadcast signals. The entrepreneurs who developed these early systems simply erected broadcasting antennas at the top of a valley and connected cables from these antennas down to the homes in the valley. They then charged a monthly access fee to those households that were "cabled." This was quite ingenious and creative for its time because it provided opportunities for people to receive television signals who ordinarily could not be able receive them.

It was also discovered that cable not only provided service, but also provided vastly improved screen-image quality as compared to broadcast images. Between 1950 and 1965, additional systems were designed and installed to improve reception of local and nearby broadcast television signals. The sometimes shaky and unclear television images

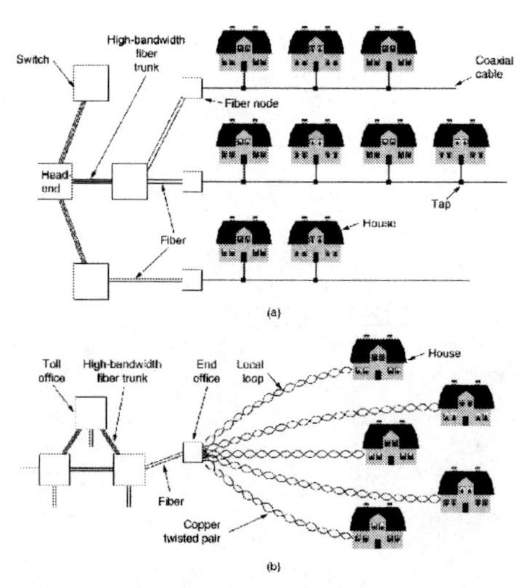

Wiring for cable communication. (Penguin.dcs.bbk)

were gone. Hence, even those receiving free broadcast signals found it worthwhile to pay a monthly fee to receive better television reception.

By 1965 there were 1,570 cable television systems (known as CATV) that provided service to 1,575,000 subscribers, or three percent of the homes in the United States. By then, cable companies saw the value of producing their own programming while also "pulling" programs from broadcast signals. Because of this, broadcasters and program copyright owners began seeing cable TV companies as a potential competitive threat. In response to pressure imposed on the Federal Communication Commission (FCC) by the "big three" broadcasters—ABC, CBS, and NBC—the FCC began to regulate cable TV by "freezing" further development in the nation's top 100 television markets until it could assess the economic impact that cable would have on the major broadcast networks.

By 1972 it became apparent that one of the major advantages of cable television is its service to local communities and that the systems, for the most part housed locally, provided opportunities for communication within communities that broadcasting did not permit. Seeing that this added benefits as a community service, the FCC began requiring cable operators to provide free public access channels. This meant that if someone in a local community wanted to develop programming that had a specific interest to those in the community, there had to be a provision made for this, and at no cost. An example of such programming would be the televising, via cable television, of local City Council meetings. Also by 1972, the number of cable systems in the United States had grown to 2,991 systems serving 7.3 million subscribers, or 11 percent of the homes in the nation.

By 1975 there was the rise of multiple systems operators (MSO). These were large cable companies that started buying up local cable systems and while also installing their own in communities not yet "wired" for cable. Hence, while on the surface it appeared that cable systems were local in structure, they rapidly became part of "big business" that saw opportunities to generate substantial revenues from cable technology, and particularly when linked with satellite transmission; an opportunity that developed in the early 1980s. This is when "pay TV" developed, as did Home Box Office (HBO), which was introduced by Time, Inc. The cable companies also saw opportunities in media other than television and began participating in mergers, joint ventures, and even acquisitions of newspapers and magazines. It was imminent at that time that cable technology was going to provide opportunities for the printing industry in producing products faster and better.

By 1977 it was clear that cable communication was not primarily for the few who could not access broadcast signals, nor was it a passing technology fad. In a landmark decision (*FCC v. HBO*) a ruling was handed down that pay TV does not threaten over-the-air broadcast television, and that the FCC "freeze" and restriction of "recent" movies on cable TV is a violation of the First Amendment. As a result the "freeze" imposed in the mid 1960s was lifted, making way for cable communication in every market in the United States in the areas of television, movies, and even printing.

Shortly after this ruling, Turner Broadcasting introduced the Cable News Network (CNN), United Press International (UPI) announced 24-hour satellite news, VIACOM International announced Showtime, and Warner Communication announced its "QUBE" experiment. The "QUBE" experiment was one of the first attempts to initiate two-way communication between the cable television company and the viewer. In other words, viewers were able to communicate back and forth with the programming they were watching with the use of a keypad in the form of a cube connected to their television sets. The system was crude in its initial implementation, but it was a clear sign of the capabilities that were to be developed in the years to come as computer technology developed into a means for people and companies to communicate with each other.

In 1978, through pressure from the cable companies, in a case involving *Midwest Video v. FCC*, the courts struck down the earlier ruling requiring cable companies to provide free public access facilities. However, the way local municipalities got around this was by specifying the free public access desire in the Requests for Proposals (RFP) issued to cable bidders for lucrative franchises. In many cases, the bidder that offered free public access was the cable company that was awarded the franchise.

By 1979 there were more than 2,000 cable systems providing at least one pay service generating approximately one-half billion dollars in annual revenue. By 1980, there were 4,600 cable TV systems serving 18,300,000 subscribers, or 21 percent of the nation's households—with the expectation of doubling in the five years to follow. By 1981 the cable systems had 107-channel capacity after previous maximums of 12, 36, and 54-channel capacity.

Since the early 1980s, the cable industry has continued to grow, with its focus on promoting pay TV programming along with other income-generating services that crossed the movie industry and print media. The rise of the Internet provided yet additional opportunities for cable companies in offering broadband services, particularly in areas where DSL is not available. Nearly every major cable television company now offers broadband services.

Satellite Communication

The first testing of communication satellites took place in the late 1950s and early 1960s. The first practical application was the launch of Syncom II in 1963 when the satellite was placed in the "geostationary" orbit 22,300 miles above the Earth's equator. This orbit is explained in detail later in this chapter.

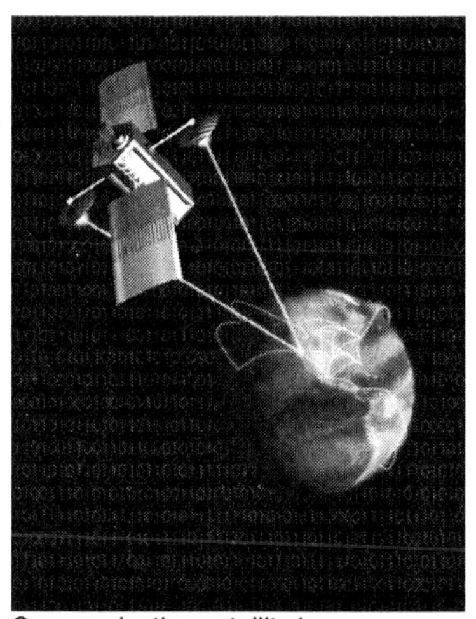

Communication satellite in geostationary orbit. (ESA Telecommunications)

By the early 1970s it was evident that satellites were going to provide enormous opportunities for advances in communication in all walks of life, and the FCC instituted the "open skies" policy. This permitted any qualified applicant to own and operate satellites in a competitive environment. By 1973 non-governmental satellites were being placed in orbit, and private companies also began renting transponder space on government-operated satellites. For example, RCA Global Communications leased capacity on a Canadian ANIK satellite to provide communication service between Alaska and the West Coast.

In 1974 Western Union's WESTAR I was placed in orbit, and between 1974 and 1980 five additional commercial satellites were placed in orbit. Also around 1975, "low cost" private receive-only earth stations became available for fixed satellite services. They used 10-meter antennas and cost between $100,000 and $150,000.

By 1976 the FCC ruled that it would approve applications for earth stations using 4.5 meter antennas or larger. These smaller "dishes" and related earth station receivers cost between $25,000 and $70,000. By 1979 the earth station costs dropped to from $10,000 to $15,000, and some were even less expensive. It was also at this time that the FCC deregulated receive-only terminals, after previously requiring frequency coordination or clearance, construction permits, and licenses. It was clear that the federal government was encouraging satellite communication by making it easier and less expensive for the private sector to participate.

By 1980 there were approximately 2,500 domestic earth stations in operation with the anticipation that this number would double in the next few years. A range of commercial and non-commercial organizations were using three domestic systems, comprising eight satellites, to distribute video, voice, facsimile, and data communication.

Since the early 1980s, well over 120 communication satellites have been placed in the geostationary orbit to handle domestic, international, and military communication, as well as scientific, meteorological, and experimental functions. Some were used for sending and receiving data for the printing industry. Commercial applications have been widespread. As was the case with cable communication, satellite transmission has provided new options for broadband Internet communication.

Satellite Transmission Today

Today, satellites offer many differing abilities depending on what functions they were designed to perform. Communication satellites generally provide telephone, television, and data services between widely separated fixed locations—for example, the switching offices of two different national telephone

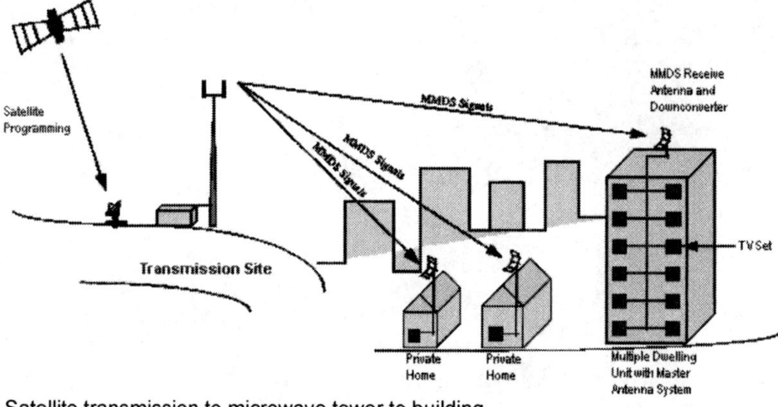

Satellite transmission to microwave tower to building. (Teleset)

networks like AT&T and MCI. Satellites also allow communication between fixed locations and mobile users such as shore stations and ocean-going ships, and between mobile users, airplanes, and satellite-based cellular phones.

Communication satellites are used in the graphic communication industry for sending and receiving data used for printing. These satellites differ from those used for weather tracking and for surveillance or other military and security purposes. Communication satellites are multiple antennas that serve as microwave relay links in outer space. These satellites are placed in the geostationary or geosynchronous orbit (also known as the Clarke Orbit, named for the person who discovered it) located 22,300 miles above the Earth's equator. This is the orbit in which satellites travel at the same surface speed that the Earth turns. Therefore, the satellite always appears in the same spot. If placed in an orbit below the geostationary orbit, a satellite would travel faster than the surface speed of the Earth. If placed in a higher orbit, a satellite would travel slower than the Earth's surface speed.

A major organization involved in the extensive use of satellite transmission is INTELSAT (International Telecommunications Satellite Organization). INTELSAT was established in 1964 by 11 nations and presently controls satellite use by more than 120 nations. INTELSAT is the world's largest commercial satellite communications provider. It has more than 20 satellites in the geostationary orbit and has access to digital channels that gather information and news found on the transponder of satellites. INTELSAT has also been providing direct-to-home satellite communication since 1992.

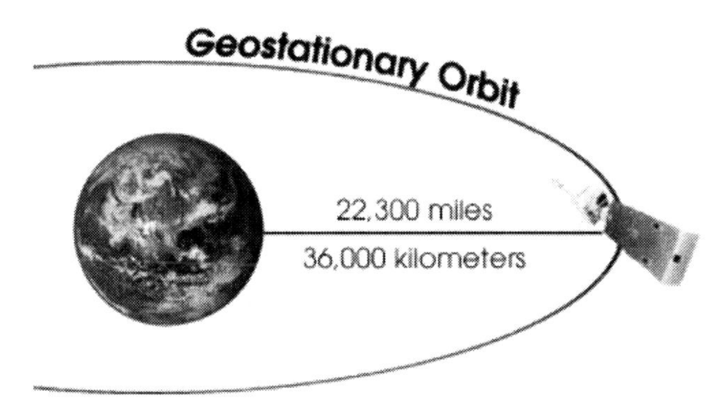

The geostationary orbit is the orbit in which a satellite circles the Earth every 24 hour or at the same rate that the Earth rotates. (Univ. of Wisconsin – Madison)

INTELSAT satellites are being used by the printing industry, as well as other industries, for receiving and sending data for printing periodicals, books, magazines, and for commercial printing. The printing industry is responsible for offering news and related services to a widely diverse group of individuals and businesses. Hence, organizations such as Dow Jones, The New York Times, Gannet Newspapers, Time Inc., and others have adopted satellite transmission for producing their publications at multiple sites around the world simultaneously. *The Wall Street Journal*, published by Dow Jones, was the first to use satellite transmission for the production of printing and, as a result, *The Wall Street Journal* became the first national newspaper produced using this technology. Gannett Newspapers was the first to use satellites for the transmission of color data in the production of *USA Today*.

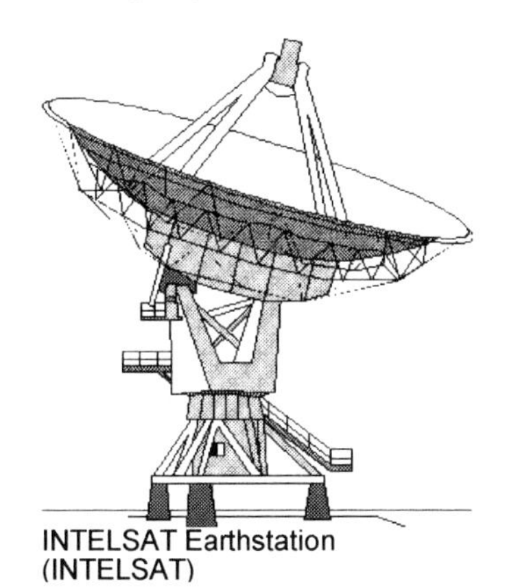

INTELSAT Earthstation (INTELSAT)

Satellite transmission basically involves transmitting signals from an earth station to a satellite. Equipment on board the satellite receives the signals, amplifies them, and transmits them to a region of the Earth. Receiving stations within this region pickup the signals, thus providing the communications link.

The main advantage of satellite transmission is the extreme bandwidth available because of the use of microwave radio, most commonly used in the super-high-frequency band. This band is known as the C-Band that specifically is the 3.7 to 4.2 gigahertz, or billion signal cycles per second, (GHz) communication band used for downlink frequencies in tandem with the 5.925 to 6.425 GHz band for the uplink. Signals operating above 3 GHz are known as microwaves; above 30 GHz they are millimeter waves. As one moves above the millimeter waves, signals begin to take on the characteristics of light-waves. These frequencies correspond to wavelengths ranging from 10 cm to 1 cm (4 inches to 0.4 inch). Such short radio waves diverge along straight lines in narrow beams, rather than propagating in an expanding spherical wave front as in the case of radio or television transmissions that use longer wavelengths. In order to communicate via microwave radio, therefore, transmitters and receivers must be situated within line of sight of one another. On land, this can be achieved by using towers or hilltop locations, but microwave communication across oceans is impossible without the use of satellites.

In order to relay signals in these frequencies, satellites rely on transponders. Transponders are the areas on a satellite required for a single channel's programming. This is a combination receiver, frequency converter, and transmitter package. Transponders have a typical output of five to ten watts. They operate over a frequency band with a 36 to 72 megahertz bandwidth in the L, C, Ku, and sometimes Ka bands—or, in effect, typically in the microwave spectrum, except for mobile satellite communication. Communication satellites typically have between 12 and 24 on-board transponders. Fulltime transponders are used for networks such as CNN that provide continuous, coverage 24 hours per day.

The most common source of microwave power for transmitting signals from communication satellites is the traveling-wave tube amplifier, the only remaining representative of vacuum-tube technology in satellites. Solid-state power amplifiers present an economical alternative mainly for lower power transmissions. Solar cells are the universal source of electric power in operational satellites. The cells can be placed on flat panels that radiate outward from the body of the satellite, or they can cover the satellite's cylindrical surface. Power is stored in nickel-cadmium or nickel-hydrogen batteries.

Major components in satellite communication that are needed to produce products of the graphic communication industry are transponders, earth terminals, and multiple-access communications equipment.

Transponders usually have separate frequency channels that operate with multiple carriers from different earth terminals to separate carrier channels. They provide isolation between high-level and low-level output. The earth terminals are designed for multi-carrier operation, which allows the product information or services to be sent to multiple sites. These can be companies or individual residences. The proper configuration of each of these components of the satellite enables the production of newspapers, books, and magazines of high quality, particularly when interfaced with desktop computers and digital printing.

The maintenance of the equipment and the continual replacement of the satellites as their orbits decay are the areas of concern. Satellites have a life span of approximately 16 years before the Earth's gravity causes them to burn up in the upper atmosphere.

Today's communication satellites typically have 24 transponders sectioned by 24 hours. Hence, a user can rent an entire transponder or a portion of one based on number of hours. Therefore, multiple users can share transponder time and, thus, share the cost. The transponder is the part of the satellite that receives and sends signals to an earth station. The earth station is conspicuous by its concave dish antenna. In the days of the large dishes of 10 feet or more, the standard was that it required three satellites to communicate around the world—one above North America, one above Europe, and one above Asia. Each satellite would "see" one-third of the Earth's arc. When a signal was sent to one, the satellite would then send that signal to the other two to complete the around-the-world communication. Today, with the use of much smaller dishes, the standard is four satellites to "footprint" the Earth, with each "seeing" one-fourth of the Earth's arc. This additional satellite also helped to improve performance at both ends of the arc that would sometimes suffer.

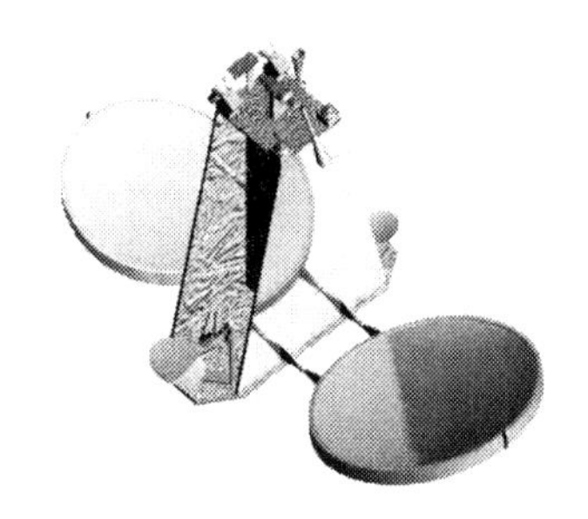

Satellite transponders
(Platform Express-1000)

In the United States, many communication satellites are placed above Denver, Colorado because Denver is considered the approximate geographic center of the nation. When transmitted from that position there is a fairly good distribution of signals to the remainder of the nation. This is also why some of the nation's largest users of communication satellites, e.g., cable television companies, are located in or near Denver.

The early users of satellite transmission in the graphic arts were the companies previously noted: Dow Jones & Co. for *The Wall Street Journal, The New York*

Times for its national edition, Gannett Newspapers for *USA Today*, Time Inc. for the production of *Time* magazine, and Newsweek for the production of *Newsweek* magazine.

The satellite transmission industry is diversified, due to its varied applications. There are hundreds of companies currently using the existing satellite networks, with more joining every day. From the military use to the home, the benefits of satellite transmission are realized throughout the world. Satellites can reach anyone, anywhere, as long as they have a receiver.

Graphic communication is a vital growing industry that touches the lives of nearly all individuals. Satellite transmission is a process that has become essential to graphic communication and its many printing industry segments. It is a link between people and technology that furthers the free flow of information.

Although there are multiple users of satellite transmission, the focus in graphic communication is on the printing industry. New printing industry technology relies on the latest technologies used for telephones, television, data transmission, and overall global connectivity—all of which has applied satellite transmission for sending and receiving. Furthermore, the Internet and the World Wide Web are now linked to these technologies, allowing the flow of information directly to and from the home or business.

Satellite transmission used in the printing industry can serve numerous purposes. Since companies may have access to multiple forms of information via satellites in the geosynchronous orbit, there may also be a carrier that enables them to perform multiple tasks. The printing plant and press department can use satellite transmission in a variety of ways to receive services, which can cover many facets of production.

Some of the major producers of digital printing presses make use of satellite transmission to have their technology perform tasks such as transferring electronic files and putting together plates containing electrophotographic images, and in the production of full-color commercial printing. In the future, electronic printing will increasingly engage the use of satellite transmission to accomplish a more global and efficient form of communication.

In an age where the speed of delivery of content for print, as well as the delivery of the printed product, has become important in the competitive media business, satellite transmission has had a positive impact on the graphic communication industry. In fact, the United States has an International Communication Policy designed to encourage and enhance the use telecommunication technology.

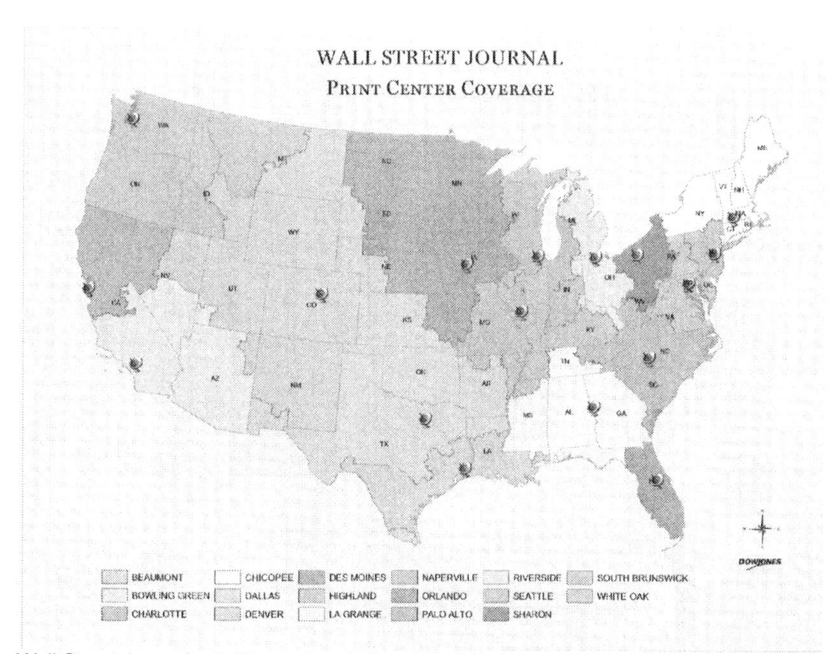

Wall Street Journal satellite centers in the United States.
(Dow Jones)

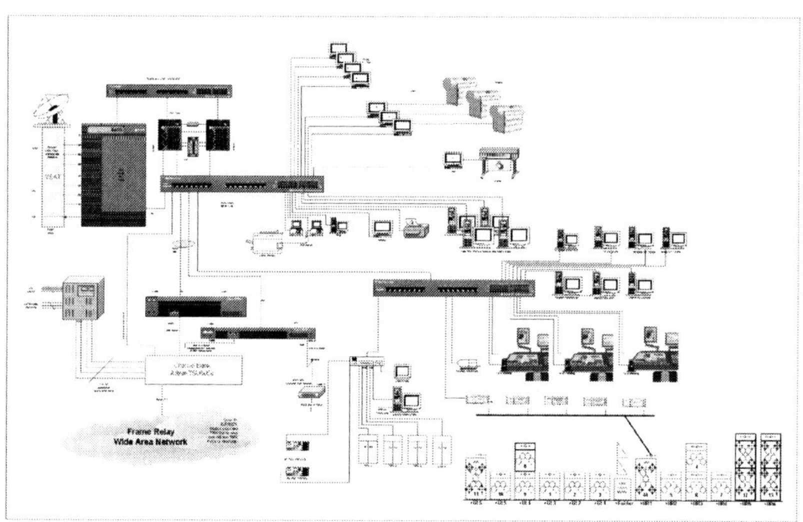

Wall Street Journal plant Network. From satellite to printing press.
(Dow Jones)

International Communication Policy

In October 1980, the United States government established an international communication policy to address the rapid developments that were taking place in communication technology. It was recognized that in the coming decades more attention would have to be paid the worldwide issues emerging from the interrelationships of computers, satellites, communications, and information.

The merging of information, computers, and telecommunications technology was opening vast new opportunities. The trends of the "information age" made it clear that the ability had to be developed within private commerce to move massive amounts of information across oceans and continents in a matter of minutes or seconds.

Satellite transmission was once completely controlled by the federal government. However, in the early 1980s the government saw the value of satellite transmission to support society's demand for vast increases in the speed, production, and distribution of information. This was a period characterized by an "information explosion." Hence, the government further opened the airwaves for commercial purposes. This has led to the widespread use of satellite transmission today in the dissemination and receipt of information. The graphic communication industry has benefited tremendously from this.

Today, using cables and satellites, the transmission of large amounts of data is easy and commonplace. Entire newspapers and magazines are transmitted throughout the country and the world quickly and without errors. The Internet has linked people together in a way never before thought possible, and by combining this with cable and satellite network, so much more is possible.

It all started with the invention of the transistor in 1947. This began a chain of communications and information technology breakthroughs that has transformed the international telecommunications landscape. The interaction of computer and telephone technologies during this time, for example, has spawned an array of innovative telecommunications services that has radically altered the way business is conducted and the way companies and individuals interact. Economic and cultural opportunities were provided on an international scale.

As a result of decisions made by Congress in the early 1980s, today the United State is guided by five fundamental communications policies.

<u>Free flow of information</u>

Congress and the Federal Communication Commission (FCC) decided that open communication and access to the airwaves by private corporations and citizens is a good thing. It was determined that the benefits that satellite systems made possible stemmed largely from a vastly increased capacity to exchange information inexpensively and reliably with all parts of the world. The ultimate result was the facilitation of world trade, education, entertainment, and many kinds of professional, political, and personal discourse that are essential to healthy human relationships and international communication. Better and less expensive communications were deemed vital elements in the growth of civilization.

Privacy

In an increasingly automated world in which large institutions, including government, hold vast amounts of data on individuals, it is important that fair information practices be followed with respect to that data. The protection of privacy has developed into a major issue in light of the growth of the Internet and the increased pressure for people and companies to transmit personal and confidential information such as Social Security numbers and credit card numbers in doing business. The extent to which people and companies avail themselves of the benefits of electronic communication largely depends on the assurance of privacy and security.

Free market forces

Since the early 1980s, the United States government moved quickly to minimize or eliminate regulations in the domestic telecommunications arena, while avoiding the imposition of regulation in the information and computer industries. The general approach to these issues is influenced by the broad trend toward reduced government regulation of economic activity. The government determined that regulation is not always the answer to furthering the use of technology, as it has often been an impediment to economic progress.

Indeed, the events of September 11, 2001, in New York City have led the government to take a close look the extent to which access to electronic communication channels should be unmonitored. However, in the give and take between stringent government control and unmonitored open access, there has to be an acceptable middle ground that protects national security and private enterprise while giving individual citizens access to communication for business, personal communication, and intellectual growth.

Free trade

An increasing number of internationally traded goods and services contain an information component, and artificial restraints on information may have more and father-reaching negative effects than are immediately obvious. The government came to realize that free trade with minimal restraint must be permitted to the extent possible, and particularly with allied nations. Such thinking has led to NAFTA and is likely to be more broadly applied in the future.

Availability of telecommunications to the public

Telecommunications and information technologies are merging in such a profound manner that it is difficult or impossible to distinguish between them in many areas of application.

The United States strongly supports the introduction of new communications carriers and new services into the marketplace, as well as the maintenance of customer choice. The overriding premise is that telecommunications should be available to all.

12

Printing Industry Business Practices

The term "business practices" as presented here refers to the common practices of the printing industry. Business practices are recommended guidelines and are flexible enough to accommodate various client and service provider relationships. These business practices are generally recognized in credible industry publications and endorsed by leading industry professional associations including the National Association for Printing Leadership (NAPL), the Graphic Arts Technical Foundation (GATF), and Printing Industries of America (PIA).

It is recognized in today's competitive environment that the most successful printers and print brokers offer their clients a full range of products, exceptional support services, first-rate customer service, timely delivery, and quality printed products.

The relationship between the printer or print broker and the client cannot be at odds. Teamwork is key to keeping customers happy as cycle times are reduced and technology constantly shifts. Achieving this requires a commitment from each team member to work toward a unified group of objectives and dedication to supporting each other's actions.

The print buyer is not always familiar with all facets of the printing process and printing production, and therefore places her or his trust in the printer or broker to provide fair and honest assessments of what it will take and cost to produce and deliver a job that meets the requirements.

Primary Duties of a Printer or Print Broker and Customer Services Representatives (CSR)

Printers and print brokers typically rely on a customer service representative (CSR) to coordinate details of work-in-process. The CSR functions as a liaison between the printing company and customer. The CSR's job involves coordinating activities between estimators, sales representatives, production management, and other internal printing company staff. It also involves facilitating timely delivery of each customer's job establishing and maintaining an agreed-upon level of quality. The CSR provides backup support for sales, estimating, and production management, and serves as the "face" of the company. In this role, the printer or print broker is expected to provide the printed job at the highest quality, as quickly as possible, and at the lowest price—with minimal to no hassle to the print buyer.

Job Specifications

In printing, job specifications—often referred to as "specs"— are the key parameters that define an order. The printer or print broker is expected to assist the client in developing proper and accurate job specifications. Accuracy of specifications is very important, as this is what price quotations are based on. The service provider can re-quote a job at the time of submission if any of the specs change or if copy, film, tapes, disks, or other input materials do not conform to the information on which the original quotation was based.

It is also important that the service provider (such as the printer or print broker) work with the client to prepare a professional Job Specification Form (JSF). This is a form that the printer or sales rep completes in discussion with the client. The purpose of this form is to provide essential information that describes the job accurately and fully. From this information the job can be planned and economically evaluated.

The Job Specification Form usually includes the following information:

• The customer's name, address, and telephone number
• The sales representative's name
• A general description or the job
• Quantity levels
• Flat sheet and folded dimensions (with possible alternate sizes)
• Bleeds and margin information
• The number of colors
• Complete details regarding all artwork (who will supply it and whether it will be ready for camerawork, scanning, or digital input)
• Complete paper specifications
• Binding, finishing, and other post press requirements
• The type of packaging to be used
• The type of delivery method and time of delivery

Some customers provide their own written specifications. However, these documents are often not thorough or detailed enough because the print buyer is not always aware of the types of details that should be included. That is way in such cases it is ethical on the part of the printing company or print broker to help the client develop specific and accurate job specifications.

The printing company or print broker should make sure that the client is not rushed in developing job specifications. Rushed or incomplete specifications typically result in costly changes and disappointments later on.

The Estimate

An estimate is generated and used by the printing company as an internal dollar measurement of the cost required to produce a given product. A cost estimate is an approximation and is generally not shared with the customer. It does not necessarily represent the final cost.

The Proposal
A proposal is a tentative offer to produce a printed product for a specific dollar amount. It too is not the final cost and is subject to negotiation and change. The proposal is a formal, typewritten document that specifies price for a printing order. A complete proposal describes the work to be done, the expected levels of quality and quantity, and provides tentative selling prices.

The Quotation
A quotation is an offer from a print broker to a customer to produce a printed product for a specified dollar amount. Unlike a proposal, the quotation is final and binding to both parties. Once the price quotation has been agreed upon, it has the impact of a legal agreement.

Because the quotation is a legal contract, prudent printers and print brokers include with it a copy of the latest Graphic Communications Business Practices (GCBP). These practices were research and drafted by the Graphic Arts Technical Foundation (GATF), the National Association of Printing Leadership (NAPL), and the Printing Industries of America (PIA) through surveys and feedback from hundreds of printers.

Typically, a quotation not accepted within thirty days may be changed.

The Order
Acceptance of orders is subject to credit approval and contingencies such as fire, water, strikes, theft, vandalism, acts of God, and other causes beyond the service provider's control. Credit approval documents are ordinarily separate from customer agreements to acquire services. Canceled orders require compensations for incurred costs and related obligations.

Alterations/Corrections and Reprints
Customer alterations include all work performed in addition to the original specifications. All such work is subject to additional charges at the service provider's current rates.

When a customer requests a straight reprint order, there is no rework necessary. Therefore, the manufacturing costs of the job are lower than in the initial run because prepress manufacturing has been completed. When a customer requests a minimum change reprint, the estimator must determine the quantity of rework necessary, then add its costs to the existing press and finishing costs. In either situation, because all or most of the prepress work has been completed for the reprint and is paid for by the customer on the first printing, manufacturing costs are reduced.

With this in mind, printing companies or print brokers provide a reprint discount on the billed price of the first printing. Discount rates and discounting vary. However, a range of 5 percent to 30 percent off of the initial price of the first job is common.

Over-runs or Under-runs

Over-runs or under-runs usually do not exceed 10 percent of the quantity ordered. The service provider will bill for actual quantity delivered within this tolerance. If the customer requires a guaranteed quantity, the percentage of tolerance must be stated at the time of quotation.

Production Planning

Production planning, or job planning, is the evaluation of manufacturing methods to ensure that the customer's order is processed to meet her or his requirements. For planning to be thorough, it should be done carefully and it should be based on the set of accurate and complete job specifications provided by the customer.

Production Schedules

Production schedules will be established and followed by the customer and the service provider. In the event that the customer does not adhere to a production schedule, delivery dates will be subject to renegotiation. There will be no liability or penalty for delays due to state of war, riot, civil disorder, fire, strikes, accidents, action of government or civil authority, acts of God, or other causes beyond the control of the provider. In such cases, schedules will be extended by an amount of time equal to the delay incurred.

Customer Credit

A customer's credit should be checked prior to extending a proposal or quotation. However, this is not always possible in an environment of quick turnaround. In such cases a statement such as "Prices quoted are conditional upon company approval of customer's credit" on the proposal or quotation clarifies the company's credit checking process. Credit check documents and quotations or proposals are separate documents; they should not be one and the same document.

There are three ways to check a potential customer's credit. One is to have the customer complete a credit application that requests credit references. A second way is to contact a credit association. The third way is to use the credit referral service offered through most Printing Industry of America (PIA) regional offices.

Markup
The most frequent "markup over cost" used by commercial printers is 35 percent. Thus, a job with a total cost of $100 would be priced at $135. However, print broker markups can range from 3 to 35 percent.

Invoicing
The customer should be invoiced as soon as possible after delivery of the printed goods. One working day or sooner is recommended; on the day the job is shipped is ideal. Thus the customer receives the goods and immediately thereafter the invoice (with payment due within 30 days). Discounting for early payment is sometimes offered—i.e., 2% net 10 day.

Experimental Work
Experimental or preliminary work performed at a customer's request will be charged to the customer at the service provider's current rates. This work cannot be used without the service provider's written consent.

Customer's Property
The service provider will only maintain fire and extended insurance coverage on property belonging to the customer while the property is in the service provider's possession, and liability for this property will not exceed the amount recoverable from the insurance. Additional insurance coverage may be obtained if it is requested in writing, and if the premium is paid to the provider.

Creative Work
Sketches, copy, layouts, comprehensives, and all other creative work developed or furnished by the customer is the customer's exclusive property. If the service provider develops such work, it remains the property of the service provider, and the creator must give written approval for any and all use of this work and for any derivation of ideas from it. In sum, the use of artwork, type, plates, negatives, positives, tapes, disks, and all other items remains the exclusive property of the provider of such items.

Electronic Manuscript or Image
It is the customer's responsibility to maintain a copy of the original file. The service provider is not responsible for accidental damage to media supplied by the customer or for the accuracy of furnished input or final output. Until the provider can evaluate digital input, no claims or promises are made about the provider's ability to work with jobs submitted in digital format, and no liability is assumed for problems that may arise. Any additional translating, editing, or programming needed to use customer-supplied files will be charged at prevailing rates.

Customer-Furnished Materials and Purchases

Material furnished by customers or their suppliers should be verified by delivery tickets. The provider bears no responsibility for discrepancies between delivery tickets and actual counts. It is up to the printer or print broker to make sure that all materials and supplies received are accurately documented. Customer-supplied paper must be delivered according to specifications furnished by the service provider. These specifications will include correct weight, thickness, pick resistance, and other technical requirements.

Artwork, film, color separations, special dyes, tapes, disks, or other materials furnished by the customer must be usable by the provider without alteration or repair. Items not meeting this requirement will be repaired by the customer, or by the provider for an additional charge. Unless otherwise agreed in writing, all outside purchases as requested or authorized by the customer are chargeable.

Prepress Proofs

The service provider will submit prepress proofs along with original copy for the customer's review and approval. Corrections will be returned to the service provider on a "master set" marked "Okay," "Okay with corrections," or "Revised proof required," and signed by the customer. Until the master set is received, no additional work will be performed.

The service provider will not be responsible for undetected production errors if proofs are not required by the customer, the work is printed per the customer's okay, or if requests for changes are communicated verbally.

Press Proofs

Press proofs will not be furnished unless they were required in the quotation. A press sheet can be submitted for the customer's approval as long as the customer is present at the press during makeready. Any press time lost or alterations/corrections made because of the customer's delay or change of mind will be charged at the service provider's current rates.

Color Proofing

Because of differences in equipment, paper, inks, and other conditions between color proofing and production press department operations, a reasonable variation in color between color proofs and the completed job is to be expected. The expected variation should be explained to the customer and the customer should acknowledge an understanding of this. Variations of this kind are considered acceptable.

Delivery

Unless otherwise specified, the price quoted is for a single shipment, without storage. Proposals are based on continuous and uninterrupted delivery of the complete order. If the specifications state otherwise, the provider will charge accordingly at current rates. Charges for delivery of materials and supplies from the customer to the service provider, or from the customer's supplier to the provider, are *not included* in quotations unless specified. However, the customer should be made aware of the approximate cost once quantities and other job specifications have been determined. Also, once quantities and other job specifications have been determined, the service provider should make the customer aware of expected delivery charges after the job is completed.

Title for finished work passes to the customer upon delivery to the carrier at shipping point; or upon mailing of invoices for the finished work or its segments, whichever occurs first.

Storage

The service provider will retain intermediate materials until the customer has accepted the related end product. If requested by the customer, intermediate materials will be stored for an additional period at additional charge. The provider is not liable for any loss or damage to stored material beyond what is recoverable by the provider's insurance coverage.

Taxes

All amounts due for taxes and assessments will be added to the customer's invoice. Tax exemptions won't be granted unless the customer's "Exemption Certificate" (or other official proof of exemption) accompanies the purchase order.

Telecommunications

Unless otherwise agreed, the customer pays for all transmission charges, and the provider is not responsible for any errors, omissions, or extra costs resulting from faults in the transmission.

Terms/Claims/Liens

Payment is net cash 30 calendar days from date of invoice. Claims for defects, damages, or shortages must be made by the customer in writing no later than 10 calendar days after delivery. By accepting the job, the customer acknowledges that he or she is completely satisfied.

Liability

A Disclaimer of Express Warrantee should be provided. In this the service provider warrants that the work is as described in the purchase order.

The customer understands that all sketches, copy, dummies, and preparatory work shown to the customer are intended only to illustrate the general type and quality of the work and that they may not be intended to represent the actual work performed.

In a Disclaimer of Implied Warrantee, the service provider warrants only that the work will conform to the description contained in the purchase order.

The provider's maximum liability will not exceed the return of the amount invoiced for the work in dispute. Under no circumstances will the service provider be liable for specific, individual, or consequential damages.

Indemnification
The customer agrees to protect the service provider from economic loss and any other harmful consequences that could arise in connection with the work. This means that the customer will hold the provider harmless on any and all grounds. This will apply regardless of responsibility for negligence.

Copyrights
The customer also warrants that the subject matter to be printed is not copyrighted by a third party. Further, the customer agrees to indemnify and hold the provider harmless for all liability, damages, and attorney fees that may be incurred in any legal action connected with copyright infringement involving the work.

Personal or Economic Rights

The customer warrants that the work does not contain anything that is libelous or scandalous, or anything that threatens anyone's right to privacy or other personal or economic rights. The provider reserves the right to use her or his sole discretion in refusing to print anything that he or she deems illegal, libelous, scandalous, improper, or that infringes upon copyright law.

13

Sustainable Graphic Communication and Corporate Responsibility in Print

By Donald Carli, Contributing Author

Due to significant global, technological and market forces the demand for corporate accountability and social responsibility is growing. Population growth, globalization, the proliferation of digital information technology, increasing evidence of climate change risks caused by human activity, political instability in the developing world and the erosion of trust in big business due to ethical and legal scandals have produced a "perfect storm" of demand for corporate social responsibility, transparency and accountability in every sphere of commerce and human communication. As a result, the need for sustainable and socially responsible graphic communication technologies and processes is expected to rise.

In the years ahead graphic communication professionals will increasingly be required to:

• Develop awareness of key sustainability and corporate responsibility trends and concepts.

• Understand the relevance of sustainability to graphic communication products, processes and technologies.

• Evaluate the risks and benefits associated with sustainable print design and production methods such as lifecycle analysis and triple bottom line accounting.

• Learn where additional information about the measurement, management, reporting and verification of sustainability performance can be found.

Sustainability management is rapidly moving into the mainstream after gathering momentum over the past 20 years. A recent survey conducted by the Economist Intelligence Unit, in cooperation with Oracle Corporation, indicates a significant rise in the number of business executives and corporate investors factoring corporate responsibility (CR) into their decision-making. The survey polled 136 executives and 65 investors in October 2004. A total of 84% of executives and investors surveyed considered corporate responsibility to be a central' or 'important' consideration in investment decisions, compared with 44% who expressed that same view five years ago. Eighty-four percent of survey respondents felt CR practices could help a company's bottom line.

One telling indicator of the demand for corporate social responsibility is that over 650 organizations now publish voluntary sustainability reports in accordance with voluntary Global Reporting Initiative (GRI) guidelines. Another indicator is the UN Global Compact, which reports that more than 1,000 companies from 53 countries are now participating in voluntary Global Compact initiatives for the management and reporting of corporate social responsibility (CSR) in their annual reports. This represents an increase of 100 percent in 2002-2003 alone.

In a keynote speech presented to hundreds of global business leaders, representatives of the press and other stakeholders attending the 2004 conference of Business for Social Responsibility Xerox CEO Ann Mulcahy described her position on CSR: "We believe passionately that good citizenship is good business. It's good for our communities, good for our people, and ultimately, good for our companies. ...We are part of an ongoing experiment to demonstrate that business success and social responsibility are not mutually exclusive. In fact, we believe they are synergistic. ...Social responsibility - like every other facet of business - is a rapidly moving target, a race without a finish line. As good as any of us might think we are today, we have to be even better tomorrow. "

Most graphic communications businesses in North America have not yet been called upon to address the issue of sustainability, however, there is growing market pressure on Fortune 500 companies and large governmental and non-profit organizations to adopt voluntary codes of conduct aligned with consensus-based principles of sustainability and corporate social responsibility (CSR). Examples of voluntary codes of conduct and reporting guidelines include those developed by associations, not for profit organizations and business coalitions such as:

- Business for Social Responsibility
- The Global Environmental Management Initiative (GEMI)
- The Global Reporting Initiative
- The Institute for Sustainable Communication
- The Metafore Paper Working Group.
- The United Nations Global Compact
- The World Business Council for Sustainable Development (WBSCD)
- The World Economic Forum

In addition to investor and activist pressure for reporting, demand and action frameworks for sustainable supply chain management and procurement are arising from organizations such as the Institute for Supply Management and the Supply Chain Council.

Another indicator of how awareness of sustainability is growing is the number of "hits" or citations a search term produces using a search engine such as Google. In July of 2005 searching the phrase "Sustainability" yielded over 20 million citations versus 1.5 million two years prior. "Corporate Social Responsibility citations increased to 1.6 million from less than 180,000 two years earlier. The term "sustainable print" yielded 344 citations versus 38 citations in April of 2003, while searching "sustainable print design" yielded 9 citations vs. *none* in April of 2003.

Once sensitized to the issues, graphic arts professionals should ask essential questions, engage in dialogue, evaluate the business case, explore the use of tools and methods such as lifecycle analysis and come to see addressing the challenges of CSR and sustainability as a "crisis of opportunity." Sustainable design, production, distribution, consumption and resource recovery strategies are essential to the future of graphic communication in print and other media.

Definitions of Sustainability:

"The principle of doing business sustainably proves its worth above all in times of economic duress and structural change. An essential first step in making sense of sustainability is knowing its history and mastering the language.

-- Heidelberg Chairman and CEO Bernhard Schreier

In all likelihood you are intrigued but unfamiliar with the term sustainability. Sustainability is a broad, complex and dynamic topic for which definitions and standards are continuing to evolve. You may also be uncertain as to how sustainability or corporate responsibility is specifically related to print media and graphic communication. Other terms often associated with the concept of sustainability are: Corporate Social Responsibility (CSR), Corporate Responsibility, Corporate Citizenship and Sustainable Development.

Definitions of sustainability typically entail integrated management of the economic, environmental and social performance of business ...often referred to as managing the "triple bottom line. " Most definitions of sustainability also include voluntary management and reporting of environmental, social and financial performance based on principles that seek to exceed ethical, legal, commercial and stakeholder expectations. Sustainable business performance is also typically defined in terms of a number of guiding principles such as accountability, transparency, active engagement and open dialogue with stakeholders.

While CSR and sustainability are related terms, they are not synonymous. According to "Our Common Future" a report made to the 1987 United Nations World Commission on Environment & Development, "Sustainable development means meeting the needs of the present without sacrificing the ability of future generations to meet their own needs . . . Economics and ecology must be completely integrated in decision-making and lawmaking processes not just to protect the environment, but also to protect and promote development."

The United Nations, the World Bank and other international public sector agencies in reference to closing the economic, environmental and social gaps between developed and developing countries frequently use the term "sustainable development". Sustainability is a more general term that has been widely used by business and other organizations in developed countries to describe coordinated activities that result in the achievement of economic development, environmental health and social equity though the use of methods such as systems thinking, stakeholder dialogue, environmental management, lifecycle analysis and triple-bottom-line accounting.

The Challenges of Sustainability

"The overall challenge of sustainability is to avoid crossing irreversible thresholds that damage the life systems of Earth while creating long-term economic, political, and moral arrangements that secure the well being of present and future generations." -- David W. Orr

The business and technical challenges facing business today call for leadership, urgency and direction. However, the challenges associated with creating a sustainable future for print will require more. They will require creativity, collaboration, systems thinking, restless inquiry and thoughtful consideration of the very purpose and nature of graphic communication. In addition they will require the application of tools and methods such as ISO 14001 environmental management systems (EMS), ISO 14042 lifecycle analysis (LCA) and Global Reporting Initiative (GRI) sustainability reporting.

Major corporations are being driven to re-examine the standards of conduct and measures of performance that determine how they do business. Business leaders are confronted daily by CNN reports of crime and corruption, rising energy costs, and a loss of trust in business. In addition, there are an estimated 63 million adults in North America who are currently considered "LOHAS" Consumers. LOHAS stands for "Lifestyles of Health and Sustainability" and describes a $226.8 billion U.S. marketplace for goods and services focused on health, the environment, social justice, personal development and sustainable living.

Under pressure from investors, activists, employees and consumers, business leaders are scrutinizing the corporate social responsibility performance of their operational practices and supply chain business practices ...including what they print, how they print and how print-related products and services are valued. Examples of this can be found in the recent efforts of Time Inc. and members of the Paper Working Group as well by the Responsible Enterprise Print Initiative being led by the Institute for Sustainable Communication with the support of the American Institute of the Graphic Arts.

For large global corporations there are significant challenges associated with living up to sustainability principles. Demands for profitable revenue growth must be balanced with the need to address a significant number of risks: global warming, energy security, biodiversity loss and illegal logging to name just a few. Yet despite these challenges a significant number of business leaders now argue that the value of corporate social responsibility extends beyond philanthropy, risk management, regulatory compliance and the generation of profits.

CEO's at companies like 3M, Bertelsmann, Canon, Daimler Chrysler, Domtar, Dow Chemical, DuPont, Ford, General Electric, Hewlett Packard, International Paper, Johnson & Johnson, OCE, Mead Westvaco, Pearson PLC, Procter & Gamble, Reed-Elsevier, Stora Enso, Time Inc., Toyota, Unilever, Xerox and hundreds of other Fortune 500 companies increasing see addressing sustainability as a challenge, but they also see it as the key to top line growth, competitive differentiation and the creation of value.

Growing populations in developing countries represent significant untapped markets that can and must be served. However, these markets can only be served by re-thinking and re-inventing every aspect of business. C.K Prahalad explores the topic in his recent book "The Fortune at the Bottom of the Pyramid."

According to Prahalad: "For companies with the resources and persistence to compete at the bottom of the world economic pyramid, the prospective rewards include growth, profits, and incalculable contributions to humankind. Countries that still don't have the modern infrastructure or products to meet basic human needs are an ideal testing ground for developing environmentally sustainable technologies and products for the entire world. Furthermore, multi-national corporation (MNC) investment at the bottom of the pyramid means lifting billions of people out of poverty and desperation, averting the social decay, political chaos, terrorism, and environmental meltdown that is certain to continue if the gap between rich and poor countries continues to widen.

According to a recent report by the World Bank the Earth's population will stabilize at 10 billion by the year 2050. About half the people will live in today's developing countries. With half a century of 5% annual growth having given them incomes of $4,000 per capita -- twice the level in middle-income countries at the turn of the century --people will still aspire to higher incomes, but desperate poverty will have been vanquished. ...However, a shadow looms when we realize that this scenario would entail a 25-fold growth in output and potentially huge increases in pollution."

On a macro level, the most significant challenge facing graphic communicators, graphic arts service providers and their suppliers is charting a course for the sustainable future of print media. Key aspects of a sustainable future for print must address print's positive effects on the natural environment and the quality of human life while delivering sustainable economic growth.

On a more practical and immediate level, graphic communicators need to consider the lifecycle impacts and aspects of the materials processes and activities employed in producing printed artifacts such as this book. Where did the matter and energy in this textbook come from? Could the fiber, minerals and polymers in your hands have been produced in a more sustainable manner? Could there have been a better way, i.e. a more sustainable way, for this textbook to meet the needs it serves without compromising the ability of future generations to do the same?

The Stone Age didn't end because they ran out of stones ...it ended because tools made of bronze were more effective in meeting peoples needs than tools made of rock. Meeting the communication needs of needs of US business and consumers more sustainably will require overcoming the inertia of current practice with momentum for change. Serving the needs of the billions of people in the developing world with graphic communication products and services is once source of the required momentum and create new opportunities for sustainable growth. Both will require creating print media design, production and delivery methods based on new materials, new manufacturing processes and new approaches to "dematerializing" and "servicizing" graphic communication products.

Practical Steps Toward Sustainable Print

"Social responsibility has economic and environmental dimensions. This broad perspective is often described as a commitment to "sustainability," which has become a term-of-art for advancing economic activity while ensuring that we can sustain our activities in a sometimes fragile world without harming the future's potential. Showing respect for these consequences is no longer a fringe issue. Businesses are driving this agenda, and designers must learn to be trusted advisors on responsible communication techniques to serve clients effectively. ... It is critical to the designer, as a trusted advisor to business on communication and positioning issues and as a crafter of design artifacts, that the profession also make these issues mainstream in its thinking."

--Rick Grefe, Executive Director of the AIGA

One of the first steps taken by corporate leaders upon having a CSR epiphany is to commission the publication of a corporate social responsibility report to signal their conversion. With the erosion of trust in business brought about by the rash of recent corporate scandals the rising risks associated with business induced climate change, and other factors there has been a bull market in the number of corporate sustainability and environmental reports being published by Fortune 500 companies.

Astute graphic communication professionals with knowledge of these sustainability trends and issues at firms like FLAG and Steele Communications are increasingly being called upon to craft the content in these reports as well as their form. Prescient service provides like Anderson Cenveo printing in Los Angeles and Sandy Alexander in New Jersey are finding that their investments in environmental management system, reduction of greenhouse gases and other sustainability initiatives provide them with numerous advantages in an otherwise hypercompetitive market for print.

Until recently, the financial business case for CSR was a tough sell. However, a recent "meta study" of fifty-two research studies made over thirty years, titled "Corporate Social and Financial Performance," found that a statistically significant association between corporate social performance and financial performance exists, varying "from highly positive to modestly positive." Using empirical data from other studies, meta study authors Marc Orlitzky of the University of Sydney and Frank Schmidt and Sara Rynes from the University of Iowa found that companies' social responsibility affected their financial performances up to 83% of the time.

A second study titled "Corporate Environmental Governance," commissioned by the UK Environment Agency in November 2004, evaluated sixty research studies over the last six years. It found that 85% of the studies showed a positive correlation between environmental management and financial performance. Its conclusion was that companies with sound environmental policies and practices are highly likely to see improved financial performance.

Beyond Recycled Paper and Soy-Based Inks

One way of dealing with complexity is avoidance. Another is oversimplification. Unfortunately there is no "silver bullet" or "one size fits all" solutions for sustainable graphic communication. The specification of post-consumer recycled paper and the use of ink based on renewable resources are important steps in the right direction. However, graphic communications professionals need to avoid simplistic approaches. They need to know more, do more and demand more. It is important to specify recycled paper, but it is not enough.

Many sustainability professionals are addressing CSR through the application of Total Quality Management principles.

Sandra Waddock, Director of the Boston College Center for Corporate Accountability provides the following observation about the relationship between quality management and social responsibility: "Although acceptance by managers of quality as a business imperative was not easy to achieve, failure to pay attention to quality now can quickly contribute to business failure. We argue that a similar evolution is occurring with respect to a company's management of labor, human rights, supplier, customer, ecological and related stakeholder practices and that companies are responding by developing responsibility management systems comparable in many respects to quality management systems already in place."

While effective quality management has proven its ability to deliver significant cost savings to companies, augmenting quality management with effective CSR management has the potential to provide far greater benefits than quality management alone. The value created by effective corporate social responsibility management can far outstrip total revenues since its benefits extend beyond cost savings and increased efficiency.

Businesses are not only valued based on tangible financial assets, but also on intangible assets such as brand reputation, good will, employee knowledge and customer loyalty. It is in these areas that effective CSR management has the most profound impact. The market capitalization of corporations has increasingly come to depend on intangibles: brand image, good will, the knowledge of employees and other assets not embodied in tangible form.

According to Emiko Todoroki of The World Bank Institute, the combined value of the top 74 global brands is estimated to be $852 billion. Correspondingly, the factors employed in building a business case for CSR management are increasingly related to intangible assets and value drivers such as:

- Risks to brand reputation
- The creation of brand image value
- Employee motivation, satisfaction and retention
- Lifetime customer value

Dr. Jürgen H. Daum, Director of Program Management for mySAP Financials and author of the book "Intangible Assets and Value Creation" maintains that business success today is no longer based on production facilities, financial capital and ownership, but on invisible and untouchable values - intangible assets -, such as relationships with business partners, brands, ideas, the quality of business processes, talented employees, corporate culture, intellectual capital and innovation power. "The importance of intangible assets, the immaterial value of companies, has greatly increased – especially in the last decade. One clear indication of the trend is that the portion of a company's total market value that exceeds its book value has increased from 40 percent in the early 1980s to over 80 percent at the end of the 1990s. That means today only 20 percent of a company's market value is reflected in its accounting system.

For knowledge-based companies, such as SAP, it is often under 10 percent. And that's exactly the problem: accounting, controlling, and management instruments have not kept pace with the economic realities of the last few decades. The largest portion of companies' economic activities, with which they create value for stockholders and stakeholders, is no longer captured systematically. Accordingly, it is not transparent internally or externally, so its importance can easily be overlooked."

Whether the issue is evaluating the intangible asset value of a brand, the value of a paper choice, ink choice, or the selection of a printer, the "lifecycle" aspects and impacts associated with each choice must be evaluated against alternatives. A product or process lifecycle analysis (LCA) involves measurement and/or estimation of how much energy and raw materials are used and how much solid, liquid and gaseous waste is generated at each stage of the product's life, from the extraction of the raw materials used in its production and distribution, through to its use, possible reuse or recycling, and its eventual disposal.

The sustainability of a product or process requires an analysis of more than recycled paper content or soy ink content. How a product is printed, distributed, used and recovered is as important as the raw materials it is made of. When developing designs or advising clients, graphic arts professionals should consider the entire production process, from paper choices that consider the raw materials and energy used to make it and ship it, to choices of printing methods to distribution options that consider the use and recovery of energy and material resources and impacts on local communities.

Increasingly buyers will ask suppliers whether they can provide independently verified information about the lifecycle environmental impacts of materials and processes. Manufacturers can apply for the International Standards Organization (ISO) 14000 and 9000 series of standards. They are international benchmark for commitment to continuous improvement in environmentally responsible performance.

Graphic communication professionals should know which of their customers and suppliers are ISO 14000 qualified and they should favor vendors and suppliers that measure, manage and report on the total environmental performance of their products and services in accordance with standards such as the ISO 14042 Lifecycle Assessment model.

Manufacturers of printing equipment, paper, plastics, adhesives, ink, toner and other chemicals employed in printing and packaging recognize the value of sustainable development, yet much of what they have accomplished is not visible to customers. Ironically, some companies see little evidence of demand for sustainable solutions among printers and other buyers of their products. In part, this may be due to the complex specification and purchasing relationships that exist among designers, printers and corporate clients. In part, it may be due to lack of information, lack of awareness, lack of knowledge or apathy on the part of buyers.

To be credible, graphic communication professionals will need to learn to speak the language of sustainability and to engage vendors, suppliers, customers and other stakeholders in meaningful dialogue about the economic, environmental and social aspects and impacts associated with print solutions. Graphic communication professionals also have an obligation to themselves and to their profession to seek the knowledge and skill required to move sustainable design and production of print from the margins to the mainstream of design practice and business communications in print.

Principles of environmentally responsible print design:

- Rethink features and functions to use less material and less energy.
- Consider closed-loop lifecycles from design through production, use and recovery.
- Design for recyclability, reusability and recoverability of energy and materials.
- Seek independently verified data about environmental aspects and lifecycle impacts.
- Select materials with less impact and toxicity (via air, water and solid waste streams).
- Increase use of recycled and renewable materials.
- Optimize production techniques to eliminate scrap, error and waste.
- Select lower-impact packaging and distribution systems.
- Design for reduced energy use, water use, and waste impacts during use.
- Maximize the length of the product's useful life.
- Recover, reuse and recycle materials at end of the product's life.

Criteria to consider in selecting a printer:

- Management commitment to environmental stewardship that extends beyond legal compliance.
- All major suppliers and subcontractors are informed of the environmental policy and encouraged to adopt similar standards.
- A dedicated manager for environmental health and safety.
- Standards-based environmental and quality management systems.

• Evidence of lifecycle thinking and continuous improvement applied to key products, services offerings and business practices.

In addition to the criteria listed above, designers and specifiers should evaluate printers based on a number of other factors. The checklist above is not a set of threshold attributes for a responsible printer, although it does detail aspects of a printer's approach toward sustainable practices designers, buyer or specifier are increasingly likely to require.

Increasingly, clients may need to report information pertinent to their procurement and supply chain activities in their CSR reports and/or in their annual reports such as:

• Facilities location, orderliness, cleanliness and environmental conditions.
• Published environmental performance improvement goals & objectives.
• Quality management processes.
• Stakeholder relationship management processes.
• Raw materials lifecycle analysis data.
• Worker health and safety data.
• Fuel and energy use data.
• Water use data.
• Air emissions data.
• Solid waste recycling and disposal data.
• Toxic emissions reporting data.
• Transportation and storage of raw materials and finished goods.
• Environmental violations, fines and lawsuits.
• Community involvement and corporate philanthropic activities.
• Public disclosure and verification of performance.
• Innovative use of clean technologies and sustainable business practices.
• Environmental stewardship certifications, citations and awards.

In addition, designers, specifiers and print buyers can will increasingly be called upon to evaluate the degree to which printing companies support environmental and sustainability education, training and awareness-building initiatives with supplier, community, governmental and non-governmental organizations.

The most prevalent standard for environmental management systems in existence is the International Standards Organization (ISO) 14001 standard. It is important to note that ISO 14001 does not mandate any specific level of environmental performance or reporting. Rather, it provides a continuous improvement framework, which can be adapted on a firm-by-firm basis. Therefore, one should not assume that ISO 14001 alone is a reliable indicator of sustainable business practices.

Some may believe that the list of issues and performance factors described above is an impossible, impractical or economically infeasible threshold, however, it is relevant to note that there are many examples of large, small and medium-sized printers in Europe, Canada and the U.S. that score well in addressing all or many of these criteria.

The study of sustainability concepts like lifecycle analysis and triple bottom line analysis may be unfamiliar or challenging for many graphic communication professionals. However, there are many not-for-profit organizations, trade associations, educational institutions and community groups that are converging on the concepts of sustainability and corporate responsibility that can provide basic information and provide a basis for the development of applications and analysis specific to graphic communication. Before you embark on your journey explore the links provided at the end of this chapter.
Whether your motivation for learning about or conducting research in the field of sustainable graphic communication is based on a moral imperative, a business case, or some combination of the two, it is important that your realize that sustainability is a journey and not a destination. In the end, rising pressure for corporate social responsibility and sustainability among Fortune 500 companies and their stakeholders will afford graphic design professionals, printers and vendors with new opportunities to escape the downward spirals of commoditization and pernicious competition that they have been trapped by in recent years.

References

Adams, Michael J., David D. Faux and Lloyd J. Rieber, "Printing Technology," (Delmar Publishers, Inc., Albany, 1988), 629 pp.

Alexander, George, "Who are the Digital Printers?" Digital Publishing Solutions, March 2005.

Allenby, Braden R. and Deanna J. Rickards, eds. "The Greening of Industrial Ecosystems". Washington, D.C.: National Academy Press, 1994.

Anderson, R.: "Mid-course correction: Towards a sustainable enterprise: The Interface model." Atlanta, The Peregrinzilla Press 1998.

Anon, "Paper Knowledge," (The Mead Corporation, Chillicothe, Ohio, 1990), 474 pp.

Anon., "Print 2000," (Printing Industries of America, Inc., Arlington, Virginia, 1990), pp. x-1 - x-5.

Anon., "Digital Basics," (Mohawk Paper Mills, Cohoes, New York, 2001), 56 pp.

Anon., "Preflight: Make Your Document Ready to Fly From the Computer to the Press," http://www.techcolor.com/help/preflight.html

Anon., "Print Process Descriptions: Printing Industry Overview: Gravure," http://www.pneac.org/printprocesses/gravure/moreinfo11.cfm

Anon., "Proofing," http://www.gt.kth.se/uttryck/95/Softproof.html

Anon., "The Digital Press Revolution, DigiTal Presses: Now a Staple in the Graphics World," Digital
Output, April, 2003, http://www.digitaloutput.net/content/ContentCT.asp?P=351

Anon., "The Printing Service Specialist's Handbook and Reference Guide," (Society for Service Professionals in Printing, Alexandria, VA, 1994), pp. G.1 - G.37.

Anon., "What is a Page Description Language (PDL)?" http://www.cs.wpi.edu/~kal/elecdoc/EDpldef.html

Anon., British Printer, Oct. 1998, http://www.uidaho.edu/pd/imgsett.htm

Anon., Mohawk Papers, http://www.mohawkpaper.com/resources/html/basics/leading_presses.htm

Apps, E. A., "Ink Technology for Printers and Students," Vol. I, (Chemical Publishing, Co., Inc., New York, 1964), 256 pp.

Apps, E. A., “Ink Technology for Printers and Students,” Vol. II, (Chemical Publishing, Co., Inc., New York, 1964), 347 pp.

Apps, E. A., “Ink Technology for Printers and Students,” Vol. III, (Chemical Publishing, Co., Inc., New York, 1964), 293 pp.

Arnold, Edmond C., “Ink on Paper: A Handbook of the Graphic Arts,” Harper & row, (New York, Evanston, San Francisco, London, 1972), 374 pp.

Avery, Allen, “Remote proofing: close at hand,” American Printer, Feb 1, 2002.

Baird, Russel N., and Duncan McDonald, Ronald H. Pittman, Arthur T. Turnbull, “The Graphics of Communication (Harcourt Brace Jovanovich College Publishers, New York, etc., 1993), 410 pp.

Bann, David and John Gargan, “How to Check and Correct Color Proofs,” (North Light Books, Cincinnati, 1990), 143 pp.

Beach, Mark, Steve Shepro and Ken Russon, “Getting It Printed,” (Coast to Coast Books, Portland, 1986), 236 pp.

Bear, Jacci Howard, “Your Guide to Desktop Publishing.”

Beck, Ulrich. Risk Society: “Towards a New Modernity” Newbury Park, Cal.: Sage. 1992 [1986].

Benyus, Janine M.: “Biomimicry: Innovation Inspired by Nature” William Morrow & Co. 1997

Bisset, D. E. and C. Goodacre, H. A. Idle, R. E. Leach, and C. H. Williams, “The Printing Ink Manual,” 3rd Ed., (Van Nostrand Reinhold Co., Ltd., United Kingdom, 1979), 488 pp.

Bixler, R. & Floyd, M.: Nature is scary, disgusting, and uncomfortable. Environment and Behavior, 5(2), p. 202-247. (1997).

Blair, Raymond and Charles Shapiro, (eds.), “The Lithographers Manual,” (Graphic Arts Technical Foundation, Pittsburgh, PA, 1980), 20 chapters.

Bridg’s, “CTP Handbook for the Graphic Arts,” (IPA, South Holland, Illinois, 2000), 24 pp.

Broekhuizen, Richard J., “Graphic Communications,” (McKnight Career Publication, 1973), 380 pp.

Brown, L., Renner, M. and Flavin, C.: Vital Signs 1997-1998: The environmental trends that are shaping our future, London, Earthscan 1997

Brown, Lawrence D. and Marcus L. Caylor. "The Correlation between Corporate Governance and Company Performance." Institutional Shareholder Services, 2004.

Brundtland, G-H. (Chair): Our Common Future: Report of the World Commission on Environment and Development, Oxford, Oxford University Press, 1987.

Bruno, Michael H, "Principles of Color Proofing: A Manual of the Measurement and Control of Tone and Color Reproduction," (GAMA Communications, Salem New Hampshire, 1986), 395 pp.

Bruno, Michael H., ed., "Pocket Pal," (International Paper Co., Nashville), all editions.

Bruno, Michael, "Principles of Color Proofing," (GAMA Communications, Salem, NH, 1986), 395 pp.

Butz, Christopher, and Andreas Plattner. "Socially Responsible Investment: A statistical analysis of returns." Basel: Sarasin Sustainable Investment, January 2000.

Carruthers, Roderick W., "Why Won't You Buy Your Printing From Us? -- A study of Outside Print Buyer Attitudes," prepared for Printing Industries of Northern California, (Strategic Management Consulting, Inc., St. Albans, Vt., March, 1984).

Cavuoto, James and Stephen Beale, "Lithotronic Imaging Handbook," (Micro Publishing Press, Torrance, CA, 1990), 218 pp.

Chen, Larry. "Sustainability Investment: The Merits of Socially Responsible Investing." UBS Warburg, August 2001.

Conover, Theodore E., "Graphic Communications Today," (West Publishing Co., St. Paul, Minn., 1985), 473 pp.

Core, Erin, "Printers Weigh: Remote Proofing Options," Graphic Arts Monthly, May 2003, p. 32.

Cornell University Library/Research Department, 2000-2003, http://www.library.cornell.edu/preservation/tutorial/technical/technicalC-01.html

Cross, Lisa, "Hard Thinking About Soft Proofing," Graphic Arts Monthly, January1999.

Crow, Wendell C., "Communication Graphics," (Prentice-Hall, Englewood Cliffs, N. J., 1986), 322 pp.

Davidson, Eric D. You Can't Eat GNP: Economics as if Ecology Mattered, Perseus Publishing, 2000.

Dennis, Ervin A. and John D. Jenkins, "Comprehensive Graphic Arts," (Bobbs-Merrill Educational Publishers, Indianapolis, 1983), 605 pp.

Denton, Craig, "Graphics for Visual Communication," (Wm. C. Brown Publishers, Dubuque, Iowa, 1992), 383 pp.

DeSimone, Livio and Frank Popoff. Eco-efficiency: The business link to sustainable development. Cambridge, MA: MIT Press, 1997.

Dixon, Frank. "Financial Markets and Corporate Environmental Results." Innovest Working Paper, 2002.

Dodt, Lorette C., "Graphic Arts Production," (American Technical Publishers, Inc., Homewood, Ill., 1990), 302 pp.

Dowell, Glen, Stuart Hart, and Bernard Yeung. "Do Corporate Environmental Standards Create or Destroy Market Value?" Management Science, August 2000.

Durning, Alan. How Much is Enough?: The Consumer Society and the Future of the Earth. New York: W. W. Norton. 1992.

Econnections: Linking the environment and the economy. December 1997, Statistics Canada No. 16-505-GPE, Ottawa, Ontario.

Ekins, Paul. Economic growth and environmental sustainability: the prospects for green growth, Routledge, New York 2000.

Eldred, Nelson, "Package Printing," (Delmar Publishing Co., Inc., Plainview, N. Y., 1993), 508 pp.
Field, Gary G., "Color and Its Reproduction," (Graphic Arts Technical Foundation, Pittsburgh, 1988),
379 pp.

Farrell, Alex and Hart, Marueen. "What does sustainability really mean? The search for useful indicators." Environment, Vol. 40(9), November 1998: 4-9, 26-31.

Field, Gary G., "Color Scanning and Imaging Systems," (Graphic Arts Technical Foundation, Pittsburgh, 1990), 309 pp.

Field, Gary G., "Tone and Color Correction," (Graphic Arts Technical Foundation, Pittsburgh, 1991),
168 pp.

Frank Farance, Farance/Edutool, "Standards Activities In Metadata," 1999-01, http://www.farance.com, http://edutool.com

Friedman, Milton. "The Social Responsibility of Business is to Increase its Profits" New York Times Magazine, September 13, 1970.

Heal, Geoffrey: "Valuing the Future: Economic Theory and Sustainability" Columbia University Press1998.

Henderson, H: Beyond Globalization: Shaping a Sustainable Global Economy, West Hartford, Kumarian Press 1999.

Hinderliter, Hal, "The New Contract Proof," American Printer, Feb. 1, 2003.

Hird, Kenneth F., "Offset Lithographic Technology," (The Goodheart-Willcox Co., South Holland, Ill., 1995), 720 pp.

Hoffman-Falk, Marieberthe, Ed., "Digital Prininting," (Océ Printing Systems, 2005), 432 pp.

Holliday, Charles O. Jr.; Schmidheiny, Stephan; Watts, Philip "Walking The Talk, The Business Case for Sustainable Development," Greenleaf Publishing. September 2002.

Howe, Walt, "An anecdotal history of the people and communities that brought about the Internet and the Web," September, 2004, http://www.walthowe.com/navnet/history.html

Institute for Supply Management Social Responsibility principles: http://www.ism.ws/SR/index.cfm

ISEA, AccountAbility 1000 - a foundation standard in social and ethical accounting, auditing and reporting, London, Institute of Social and Ethical Accountability 1999.

INTELSAT Satellite Services, http://www.intelsat.com

Larish, John, "Digital Photography," (Micro Publishing Press, Torrance, CA, 1992), 208 pp.

Leiner, Barry M., Vinton G. Cerf, David D. Clark, Robert E. Kahn, Leonard Kleinrock, Daniel C. Lynch, Jon Postel, Larry G. Roberts, Stephen Wolff, "A Brief History of the Internet," (Internet Society, Reston, Virginia, 2005).

Leonard-Barton, Dorothy and William A. Kraus, "Implementing New Technology," Harvard Business Review, November-December, 1985, pp. 102-110.

Levenson, Harvey R., "Electronic Digital Photography," 1994 Technology Forecast, (Graphic Arts Technical Foundation, Pittsburgh, PA, 1994), pp. 10-12.

Levenson, Harvey R., "From McLuhan to Wilkens: Bridging the Technologies of Design, Print, and Telecommunications at Cal Poly," The Prepress Bulletin, July/August, 1985.

Levenson, Harvey R., "Complete Dictionary of Graphic Arts and Desktop Publishing Terminology," (Summa Books, Thousand Oaks, Calif., 1995), 271 pp.

Levenson, Harvey Robert, "Understanding Graphic Communication," (Graphic Arts Technical Foundation GATFPress, Pittsburgh, Penna., 2000), 248 pp.

Levy, Uri and Gilles Biscos, “Nonimpact Electronic Printing,” (InterQuest, Ltd., Charlottesville, Virginia, 1993), 314 pp.

Lipetri, Joe, “Remote Proofing Delivers,” American Printer, Sept. 1, 2001.

McDonough, William; Braungart, Michael: “Cradle to Cradle: Remaking the Way We Make Things” North Point Press, 2002.

Meadows, Donella H., Dennis L. Meadows and Jorgen Randers. Beyond the limits. Toronto: McClelland & Stewart Inc., 1992.

Meadows, Donella H. et al. The limits to growth. Washington D.C.: Potomac Associates, 1972.

Miley, Michael, “The Ties that Bind,” Electronic Publishing, October, 2004.

Molla, R. K., “Electronic Color Separation,” (R. K. Printing & Publishing Company, Montgomery, West Virginia, 1988), 288 pp.

Mort, Richard A., “How To Save A Bundle on Printing,” (Richard A. Mort, Portland, Oregon, 1989), 149 pp.

NAPL- National Association of Printers & Lithographers, http://www.napl.org

Nothmann, Gerhard A., “Nonimpact Printing,” (Graphic Arts Technical Foundation, Pittsburgh, Penn., 1989), 110 pp.

Oakley, A. L. and A. C. Norris, “Page Description Languages: Development, Implementation and Standardization,” Electronic Publishing, Sept. 1988, Vol. 1(2), 79–96.

Ponting, Clive: A Green History of the World. Penguin, 1991.

Polishuk, Tom, Ed., “Combine and Conquer,” Package Printing, October 8, 2004, 3 pp.

Prust, Z. A., “Graphic Communications: The Printed Image,” (The Goodheart-Willcox Co., South Hoplland, Ill., 1989), 544 pp.

Prust, Zeke A., “Graphic Communications: The Printed Image,” (Goodheart-Willcox Co., Inc., South Holland, Illinois, 1989), 544pp.

Rea, Douglas Ford, "Electronic Still Photography," ESP '93 Teleconference Program Notes, (Rochester Institute of Technology, Rochester, N. Y., April-May, 1993), pp. 30-37.

Reid, David. Sustainable development: An introductory guide. London, UK: Earthscan Publications, Ltd., 1995.

Romano, Frank J. and Richard M. Romano, "The GATF Encyclopedia of Graphic Communications," (Graphic Arts Technical Foundation GATFPress, Sewickley, Penna., 1998), 945 pp.

Romano, Frank J., Digital Printing, (Windsor Professional Information, LLC. San Diego, 2000), 262 pp.

Segal, Ben, CERN IT-PDP-TE, "A Short History of Internet Protocols at CERN," April, 1995, http://ben.home.cern.ch/ben/TCPHIST.html

Solow, Robert M. (1991). Sustainability: An Economist's Perspective. The Eighteenth J. Seward Johnson Lecture to the Marine Policy Center, Woods Hole Oceanographic Institution, at Woods Hole, Massachusetts, on June 14. Reprinted in: Stavins, Robert N. (2000). Economics of the Environment: Selected Readings, fourth edition. W.W.Norton & Company: 131-138.

Spilker, J.J., "Digital Communications by Satellite," (Prentence Hall Inc., New Jersey, 1982).

Todoroki, Emiko: "Globalization and Corporate Social Responsibility" The World Bank Institute, May 2002.

Walker, John R., "Graphic Arts Fundamentals," (The Goodheart-Willcox Co., South Hoplland, Ill., 1980), 320 pp.

Wentzel, Fred, Ray Blair and Tom Destree, "Graphic Arts Photography: Color," (Graphic Arts Technical Foundation, Pittsburgh, 1987), 151 pp.

White, Alan T. "Putting a Price on Nature," Earthwatch Institute Journal, August 2000. Reprinted in: Ocean Seas: The Online Magazine for Sustainable Seas, October 2000, Vol. 3(10)

World Business Council for Sustainable Development. Eco-efficiency and cleaner production: Charting the course to sustainability. Geneva: WBCSD, 1994. 17 p.

World Commission on Environment and Development. Our common future. Oxford, UK: Oxford University Press, 1987.

Xerox, http://www.xerox.com/PARC/dlbx/library.html

Zeltser, Lenny, April 21, 1995, http://www.zeltser.com